MW00973987

Papers Presented to The Round Table

By Thomas J. Sherrard

Cover design and book layout by Kim Lajeunesse.

Published by NextGen Story: Custom Publishing

www.nextgenstory.com

Preface

Membership in The Round Table is comprised of 24 men and women. Consistent with the traditions of the group, an effort has been made to draw 12 of the members from the academic world (the "Gowns") and 12 members from non-academic pursuits, such as business, law, and medicine (the "Towns"). The Round Table held its inaugural meeting on February 28, 1884 and has met on the first Thursday of every month between October and June during the course of 140 years. Each member is expected to host a dinner for the group at which a member presents a paper that the member believes to be of general interest to the group. A great deal of time and effort goes into the presentation of these papers, and the objective is to inform, amuse, and provoke discussion.

In this book, I share papers written for The Round Table from 1986 to 2020. This collection includes diverse topics ranging from law and theology to science and the military. You will discover the mysteries of light and time, learn about heroic figures, delve into the workings of extraordinary clocks, and explore what makes us laugh. The collection also discusses intuitive intelligence, modern global challenges, and military resilience. I hope that this anthology offers thoughtful, balanced perspectives on topics of interest to any curious, engaged citizen.

Tom Sherrard
May 2024

Papers Presented to The Round Table

Table of Contents

The Round Table Presentation | May 1, 1986

Beyond Belief
Law And Theology In South Africa

1. Introduction

South Africa is a land of stark contrasts and ironies. Describing it often calls for the use of superlatives, both good and bad. It is the only first world nation on the entire continent of Africa. It has a standard of living for its white population that is as high as that of any nation on earth; yet that standard of living has been built and sustained with the equivalent of slave labor, and in the "homelands," separate land areas carved out for each tribal group, less than 50% of the black children live to the age of five. It is a nation that is hated and envied by the rest of Africa. No African nation, save Malawi, has diplomatic relations with South Africa, and virtually none admits to doing business with it. Yet each year at least twenty African nations indirectly support Apartheid by purchasing more than $1 billion worth of South African goods and services.

South Africa's military might is first rank. Not only has it developed a nuclear weapons capability, but its conventional forces are highly trained and well equipped. Analysts believe that the South African military could take on a combined force composed of all the armies of the nations of sub-Sahara Africa and still punch through to the Sahara Desert in approximately three weeks.

South Africa probably maintains the most complex system of human control of any nation in the world, including the Soviet Union, and this control is maintained not by brute force (though there is certainly some brutality) but by an elaborate system of laws in which there is nearly complete acquiescence by the Blacks and Coloureds to which most of the laws are directed.

South Africa purports to be a parlimentary democracy; however no Blacks or Coloureds participate in any meaningful way in the process, and, because of the unity of the Afrikaaners, the government of a nation of 30 million is controlled by less than 3,000,000 whites.

These and many other ironies tend to fascinate and disturb most outside observers, including myself. My original interest in South Africa focused upon the bizarre system of laws created by Apartheid, and my original intention was to present an analysis and perspective of some of these laws. That is still the purpose of this paper, but not the only one. My further investigation led to something which, for me, was even more curious: I found that the institution perhaps most responsible for the creation and continued justification of the doctrine of Apartheid was the Dutch Reformed Church of South Africa.

2. Geography and History

South Africa is approximately twice the size of Texas. It is comprised of four provinces, the Cape Province in the south and west, Natal on the Indian Ocean in the east, so named because it was first sighted by Vasco de Gama on Christmas day during his circumnavigation of Africa on his way to the spice islands, and the Orange Free State and the Transvaal in the central and northern interior of the country. Cape Town, overlooking the confluence of the Indian and Atlantic

Oceans, has been considered one of the most beautiful cities in the world. On Table Mountain outside of Cape Town there are more varieties of wild flowers than can be found in all the British Isles. Of the three other major cities, Johannesburg and Pretoria are located in the Transvaal, and Durban on the Indian Ocean in Natal.

The presence of whites in what is now known as South Africa began in 1652 when the Dutch East India Company established a refreshment station for its trading ships at the Cape of Good Hope. One of the myths that Apartheid has perpetuated is that at the time the Europeans occupied the Cape, the land for hundreds of miles around was empty except for a few bushmen and nomadic hunter-gatherers who became known as Hottentots because of the bizarre clicking sounds that were part of their speech. In reality, however, Black migrations into most of South Africa had commenced from the north beginning in the third century A.D. and culminating in approximately 1500. It is true, however, that the major tribes, the Zulu and the Xhosa, settled in the region known as Natal.

Soon after establishing a foothold at the Cape, the Dutch East India Company began to grant land for settlement. Over the next century settlers arrived primarily from Holland and Germany (along with a few French Huguenots) and settled the land near the Cape. By 1800 these settlements had moved out two or three hundred miles to the borders of the Namib Desert on the north and approximately the same distance to the east where Boers (the Dutch word for farmers) encountered the first significant permanent Black settlements. The farmers who participated in this migration were known as Trekboers - seminomadic farmers who lived on the land until it was used up and then moved on.

This then was the extent of white expansion until the 1800's when several events conspired to cause the Boers to expand their frontiers. First, in 1805 Great Britain conquered the

Cape Colony, garrisoned it and encouraged the settlement of the Cape by families from England. By the 1830's, England had a significant civilian and military presence at the Cape, and was slowly imposing its rule upon not just the inhabitants of Cape Town but the countryside as well. Beginning in the 1820's, the Zulu nation under a chief called Shaka, a cruel and utterly ruthless leader, waged war on the other African tribes located in the area we know as Natal, and the result was a depopulation of the Black tribes in east South Africa. In addition, in 1830 the British Parliament emancipated all slaves throughout the British empire. Slaves had been imported into Cape Town and South Africa since the 1660's, and the Boers were accustomed to having numerous slaves serve them on their farms. Therefore, pushed by a general distaste for British rule and their opposition to Parliament's emancipation of slaves, and drawn by the perceived weakness of the Black tribes in east and northeast South Africa, the Boers, in the mid-1830's, embarked on a mass migration that has come to be known as the "Great Trek."

It was during this Great Trek that an event occurred which has served to galvanize and unify the Afrikaaners. In 1838 Dingaan, Chief of the Zulu, began attacking and killing Boer families as they pushed into territories traditionally belonging to the Zulus. Ox-drawn wagon trains of Boers were attacked and wiped out by Zulu armies. One particular sizeable wagon train, traveling in northern Natal, learned that it was to be attacked by the main contingent of the Zulu army. Led by Andres Pretorius, a Boer with military training who became one of the first Boer heroes, the wagon train drew itself into a laager - which is an encirclement of wagons with the spaces between the wheels filled with dense thorny brush. With the advance warning Pretorius had an opportunity to choose his defensive position well. At the edge of a river which later became known as Blood River, he drew his wagons together and prepared for an attack which came on December 16, 1838, On that day a group of 400 Boers and Coloured slaves stood against wave after wave of attacks from more than 12,000 Zulus. When the day was over and the Zulus had withdrawn, they left more than 3,000 dead in great heaps just outside the laager, It was said that the river ran red with blood. Only two Boers were wounded and those wounds were superficial. To the Boers who took part in the Great Trek and to their descendants, the Afrikaaners of today, the battle of Blood River attained a special mythic and theological significance. It is now known as the Day of the Covenant, a day when God made a covenant with his chosen people to guide them into Canaan and assure their prosperity forever.

In the decades following the Great Trek, the Boers established two republics known as the Orange Free State and the Transvaal. At the same time, the British had taken effective control of the Cape Colony and Natal.

But the Boer republics were short-lived, The discovery of diamonds in Kimberly, Orange Free State, and of gold on the Witwatersrand in Transvaal, resulted in further British interference and pressure and ultimately an effort to control these two republics, including an abortive coup engineered by Cecil Rhodes. This pressure from the British culminated in the second Boer War which lasted from 1899 to 1902.

At the height of that War the British brought approximately 500,000 soldiers against some 80,000 Boers. The Boers fought as guerillas and ultimately caused Lord Kitchener, the British Commander, to undertake a scorched earth policy in order to subjugate them. Lord Kitchener also determined to place all Boer women and children in internment or concentration camps to prevent them from assisting the Boers' war effort. Because of disease, inadequate food supplies and sheer neglect, more than 25,000 lives were lost in these camps, far more than were lost in the fighting itself. The suffering and death of Boer women and children in these camps played a significant role in the growth of Afrikaaner nationalism following the War.

When peace finally came it was on generous terms to the Boers who lost no time in moving to unite themselves under the leadership of three men: Louis Botha, Jan Smuts and James Hertzog, all of whom had been Boer generals.

In 1910, the Cape colony, Natal, the Orange Free State and Transvaal formed the Union of South Africa and became a member of the British Commonwealth. From 1910 until 1948, the position of Prime Minister was filled by one of those three gentlemen.

The two decades following the Boer War were marked by the development of an intense nationalism on behalf of the Boers who were now calling themselves Afrikaaners. During this period, under the leadership of James Hertzog, an ever larger group of Afrikaaners sought to separate themselves from English influence and to preserve and encourage the Afrikaaner culture, religion and ideology.

Smuts, for his own part, encouraged greater ties to England and, as a consequence in 1914, he and Hertzog broke ranks with Hertzog forming the National Party and purporting to represent the pure Afrikaaner. Hertzog and Smuts reunited again in 1934 to deal with economic crises facing South Africa. It was then that Daniel Malan broke away from this coalition and formed a purified National Party with its themes of unity of all true Afrikaaners and ridding South Africa of insidious British influence. World War II played into the hands of the Malanites, for it showed Smuts as highly supportive of the English. Moreover, after the War Blacks began asserting themselves by leaving the Bantustans, or reserves set up for them, and migrating to the cities, and Smuts apparently lacked the ability or will to stop them. This allowed the Malanites, the new National Party, to play to the racial anxieties of South Africans, and a new slogan was coined for the election in 1948: "Apartheid," or "apartness." In the 1948 election, with a minority of the popular vote, the Afrikaaners, with the help of a small splinter party, won a working majority in Parliament. Since that time they have never relinquished effective control of the government of South Africa.

3. The Apparatus of Apartheid.

Promptly after their ascension to power in 1948, the National Party unveiled its program of Apartheid. This program was not created out of whole cloth. In fact, as early as 1905, the British South African Native Affairs Commission concluded that Blacks and whites should be separated in terms of politics, and land occupation and ownership. The Commission concluded that South Africa should depend on Blacks as their primary labor source. In addition, Blacks were encouraged to inhabit Bantustans that were set aside for them - much like the Indian reservations in the United States. But, until 1948 and the advent of Apartheid, there was no systematic effort undertaken to assure the segregation of the races and the total dominion and control by the whites.

The apparatus of Apartheid can generally be grouped into four distinct categories: racial classification; creation of the Homelands, or Grand Apartheid; segregation with respect to the use of amenities, or Petty Apartheid; and restraints upon civil liberties.

The keystone of Apartheid is that a person's rights and privileges in South Africa depend on one thing and one thing only - his race. The Population Registration Act of 1950 recognizes three racial classifications: white, Coloured and African. An agency known as the Race Classification Board is vested with the authority of determining in particular cases whether a person falls into one of the three racial divisions. In carrying out its statutory charge, it must rely upon the following statutory provision in order to determine whether or not a person is white:

A white is any person who in appearance obviously is, or who is generally accepted as a white person, other than a person who, although in appearance obviously a white person, is generally accepted as a Coloured person.

In determining one's classification as a white, the Race Classification Board is authorized to consider skin color and tint, location of cheekbones, hair color and texture, width of the nose, and the way a person walks. Obviously family history is also important. But many of the guidelines are so vague as to permit the most arbitrary of decisions.

As you can imagine the Population Registration Act has been a vehicle for human suffering in South Africa since it was enacted. One South African legal authority stated that "Families are torn apart when husbands and wives, parents and children, brothers and sisters are differently classified, with all the ensuing consequences to their personal, economic and political lives." Of course, classification downward not only means that one is viewed as being at an inferior social level, but in very real terms it means that living conditions will be worse, opportunities for work and leisure will be limited or eliminated and even the right of movement may be restricted. In a given year the Race Classification Board will actually reclassify as many as 1,000 people.

To insure further the purity of the racial groups created by this law, statutes prohibiting mixed marriages and interracial sex have been enacted. Although these laws have now been relaxed, one provision made it a crime punishable by up to five years in prison to engage in "any immoral or indecent act involving a European and a non-European."

If racial classification is the cornerstone of Apartheid, the "Homelands" policy represents the ultimate goal and purpose of Apartheid. The Homelands policy is expressed by means of several catch words: "self-determination;" "separate develop ment;" "multinational development." However stated, the purpose of Grand Apartheid is to locate every South African

Black in an independent Homeland. As an official explanation of this policy goes:

South Africa's objective is self-determination for all its peoples. The government's fundamental aim with self-determination . . . is the elimination of the domination of one group by another. The very purpose is to facilitate the development of each people into a self-governing national entity, cooperating with others in the political and economic spheres in a manner mutually agreed upon.

There is, of course, a more plausible reason for Grand Apartheid. It is unstated but is obvious to most who consider the facts. With less than 15% of the population being comprised of whites, whites have come to realize that their domination of Blacks in South Africa cannot prevail forever. The answer: set up separate and independent nations in which all Blacks can live. The Blacks become citizens of those nations; hence they are no longer citizens of South Africa, and South Africa no longer has to worry about their representation in the political system. Then all Blacks that come into South Africa do so as visitors or temporary workers.

To date, ten Homelands have been created. Of these, four have been officially classed as independent nations: Transkei, Bophuthatswana, Venda and Ciskei. The other six Homelands are in various stages of development. All ten Homelands, taken together comprise 13% of the land area of South Africa, a greater part of which has been determined by whites to be undesirable for use or development by them. To these tiny Homeland areas, the South African government seeks to move 75% of the population, some 23,000,000 Blacks. The Homelands are incapable of supporting themselves. In fact, many of the Homelands are not even contiguous land areas. To assure whites a continuous supply of cheap black labor, Homelands have been broken up and gerrymandered in order that a small Homeland area can be located near a city such as Johannesburg or Pretoria.

Perhaps the most famous Homeland is Bophuthatswana where Sun City is located, a Las Vegas style gambling resort. Sun City has sponsored invitational golf tournaments where the first prize is $1,000,000. It, along with the other Homelands, has also sought to attract entertainers and athletes from the United States. In most cases, these invitations have been refused on the premise that to accept would be to give the Homelands - and South Africa's scheme - credibility. You may recall that in 1981 Frank Sinatra provided Sun City and Bophuthatswana with a much needed sense of legitimacy by giving nine concerts at Sun City for a reported fee of $1.6 Million. The $85 ticket price pretty much assured that the audience would be mostly white, though Sinatra insisted throughout that he was in no way supporting segregation. "I play to all," he said, "any color, any creed, drunk or sober."

Just as the Homelands policy is an effort to move the Black population into separately designated areas, the Group Areas Act of 1950 assures that whites need never live near any Black or Coloured. This Act provides for the creation of separate group areas in all towns and cities throughout South Africa for whites, Coloureds and Africans. In practice it has allowed whites to determine where they wish to live and if any Blacks or Coloureds live in those areas, they will be forceably resettled. The most extraordinary example of application of this law occurred in Cape Town, in an area called District 6. District 6 had been an enclave of the Cape Coloureds since anyone could remember, but on its fringe lived approximately 800 whites. In 1966, at the request of the 800, District 6 was declared a white area and the 60,000 Coloureds were forceably moved to an area 10 miles outside of the city limits of Cape Town, and, as compensation for this forced taking, were paid what they claimed to be only a small portion of fair value. A recent study estimates that 3.5 million Blacks have been "removed" since 1959, most of them required to settle in the rural Homelands far from their real homes. Another 2 million more are scheduled for resettlement.

In order to disassociate further Blacks and Coloureds from whites, a series of laws have been enacted restricting the movement of Blacks in areas that are designated as white. It is a crime for a Black to be in a prescribed white area for longer than 72 hours. Even so, any Black entering a white area must be there for an authorized purpose and must depart within the 72-hour period. These restrictions on the freedom of movement of Blacks is assured by the notorious pass system, which, incidentally, is to be modified according to reports out of South Africa last week. The laws that create the pass system require that all Blacks over the age of 16 be fingerprinted and issued a reference book. A policeman may ask a Black to produce his pass book at any time, and failure to produce it on demand can be a criminal offense punishable by imprisonment up to three months.

Related to Grand Apartheid are a series of laws and regulations that seek to segregate white from Black and Coloured in everyday life. This system of laws is known as Petty Apartheid and is symbolized by the "whites only" signs that can be found all over South Africa. Of course "separate" in South Africa does not mean that the facilities must be equal or comparable for all racial groups. For example, if a Black wants to use a restroom in Johannesburg, he has a choice of 161 public toilets. All the others are reserved solely for whites.

The laws respecting Petty Apartheid are similar to the Jim Crow laws that were commonplace throughout the American south. Public facilities, buses, taxis, beaches, pools, hotels, entertainment and sports events all are subject to segregation. Joseph Lelyveld recalls a story he filed while a correspondent for the New York Times in Johannesburg. It concerned a Black choral group that had been denied permission to sing Handel's Messiah to a white audience because it had been scheduled to perform with a white orchestra. In the mind of some white official, that constituted illicit race mixing. When the sponsors of the concert proposed as a compromise that the chorus be accompanied by an organist instead of an orchestra, official permission was forthcoming, though the only available organist was also white. One of the sponsors wisecracked "We had to promise that he would play only on the white keys." As Lelyveld stated, "Apartheid in those days was often nearly funny; the reality it created seldom was."

The first three prongs of Apartheid establish a system of race classification, separate development in the Homelands, segregation in the cities and in the use of public and private facilities. In order to preserve this system a number of laws have been passed to silence its critics. These laws fall under various names: the Terrorism Act, the Unlawful Organizations Act, and the Internal Security Act. The combination of these laws and the regulations related to them provide South Africa with the legal grounds to silence any opposition. And, since judicial review of Parliamentary acts is not recognized by the courts in South Africa, there is no effective check against the breadth or whimsicality of these laws.

For example, the Internal Security Act, which allows the government to cloak its powers of censorship under its concerns for state security, makes criminal any acts or omissions that "aim at the encouragement of feelings of hostility between the European and non-European races of the Republic." The Terrorism Act prohibits any acts designed "to cause, encourage or further feelings of hostility between . . . inhabitants of the Republic or to embarrass the administration of the affairs of the State."

One of the most unusual aspects of the Internal SecurityAct is its authorization of the practice of banning. Although banning is not as severe a restriction on personal liberty as imprisonment, it has the effect of destroying any semblance of a normal existence and in some ways is more frightening than actual incarceration. A banned person is generally restricted to a certain area around that person's residence; he is prevented from entering any educational institution or publishing house or court; he is barred from attending political, social and other gatherings and is prohibited from communicating with specified persons, usually others who have been banned. Normal employment by a banned person is impractical. In its most extreme form, and the form applied until recently against Winnie Mandela, banning can amount to house arrest.

The procedures for banning are totally arbitrary. The banned person has no opportunity to contest the banning order and authorities need give no reason for their action. There is no appeal. Furthermore, the courts have no power to intervene. Banning orders apply for five years, at which time a new banning order is issued. As of mid-1980, more than 150 persons - most of whom are Black ,as well as one notable Afrikaaner clergyman who will be discussed later - were under banning orders.

Finally, freedom of association and assembly have been severely restricted. Under the Unlawful Organizations Act and the Riotous Assemblies Act, the authorities can essentially prevent public or private meetings of two or more persons. In 1976, acting pursuant to these laws, the Minister of Justice banned all outdoor meetings and processions except sports events and gatherings for which a permit was specifically granted. In 1980, in order to clamp down on political discussion during the school boycotts by Black children, the Department of Justice issued an order notable and intimidating by its breadth:

Whereas I deem it necessary for the maintenance of public peace, I hereby prohibit any gathering of a political nature at which any principle or policy or action of a government or of a political party or political group is propagated, defended, attacked, criticized or discussed except for such gatherings which I expressly authorize.

Quite obviously this decree would, if applicable here, make

it impossible to hold our meeting tonight without express approval.

Therefore, not only does the government of South Africa have in place the laws providing for complete and permanent segregation of races, but it also has a wide variety of tools to preclude any changes being made in the system it has created. The laws permit wide use of unchallengable discretionary power, leaving any opponent of Apartheid under constant threat, and keeping government officials increasingly out of touch with the level of anger and bitterness of the people affected.

4. Apartheid and the Afrikaaner Church

Now, having viewed the apparatus of Apartheid, I want to turn to one of its principal causes, and that, paradoxically enough, is the Dutch Reformed Church in South Africa. A unique aspect of the development of the culture and special identity of the Afrikaaner is the special role played by the church and its Calvinist theology. It is from this melding of Afrikaaner history and tradition with Calvinist theology that the doctrine of Apartheid was developed. The Dutch Reformed Church not only helped to create and nurture the doctrine, but continues to be one of its staunchest supporters today.

Calvinism came to South Africa in 1652 with the advent of the Dutch East India Company and the first Dutch settlers. Holland at that time served as a haven for Calvinists and other protestants fleeing from the excesses of the counter-reformation in other parts of Europe. The Dutch Reformed Church established a formal presence in South Africa in 1665 with the arrival of the first minister, or predikant, from Holland. Until 1805, when the British occupied the Cape on a permanent basis, all the ministers sent to South Africa were trained in Holland.

Even before the arrival of the first predikant, in 1658 a convoy of slaves from East Africa arrived whose purpose was to provide most of the manual labor for the Dutch settlers. The settlers also began to press into service as slaves the aboriginal Hottentots who inhabited the region. Slavery soon became a way of life for those early Cape whites, and as they moved into the interior, they, of course, took the slaves with them. According to the 1798 census in the Cape Province there were 26,000 slaves and 15,000 Hottentot servants to serve the needs of 22,000 whites.

From the diaries and early accounts of these Boer settlers, one is able to see various forces at work during the 17th and 18th Centuries. First of all, slavery was fully accepted, even taken for granted. The Boers regarded the black man as inferior in all respects. In many ways the doctrines of Calvinism supported, and even encouraged, this attitude. Out of the doctrine of the elect, from which the slaves were obviously excluded, came the assurance that God's Will decreed white mastery and superiority over these people. As the Boers settled the interior, often after killing or driving off any natives who resisted them, they took comfort from their Calvinist heritage that they. were quite properly being used as instruments in God's great plan. These Boers frequently ref erred to themselves as a chosen people, that they, like the children of Israel, were destined to occupy and exert dominion over southern Africa. The Bible, especially the Old Testament, became their guide and their defense. This attitude was echoed by an editorial in the largest Afrikaaner newspaper in 1976: "Africa's unique white tribe of Afrikaaners . . . see themselves as a sort of Israel in Africa, with a sense of God-guided destiny that it would be as perilous to discount as in the case of the original model."

This combination of forces that was beginning to mold itself into a distinct Afrikaaner identity was fostered as a result of the particular isolation of these Dutch Calvinists. Indeed, it has been suggested that isolation may be one of the most significant single factors in the development of the Afrikaaner's national character.

Until the 1800's very few ships stopped at the Cape and few visitors stayed long enough to bring the settlers news of the world, much less to enrich their lives with new ideas. In the interior the Boers were even more isolated. In fact, formal education and literacy were practically non-existent on the frontier. The image of the Boer that is historically represented is that of a farmer with a Bible in one hand and a rifle in the other. In most cases his Bible was the Boer's sole source of education and spiritual and cultural fulfillment. He was a person of predominantly rural values and a conservative, Old Testament religious heritage.

Events in the 1800's served to give added impetus to the development of the Boer's identity as a special and distinct people. The Boer resentment against all things British led first to the establishment of the Boer Republics, and, when British imperialism sought dominion over those republics, then to the Boer War at the end of the century. The victories of the Boers over the various African tribes, highlighted by the Battle of Blood River, increased the Boer's sense of destiny as a people. At the same time the Boers became totally committed to racism as a way of life. When the Republic of the Transvaal was founded in 1858, the following clause was written into its constitution: "The people are not prepared to allow any equality of the non-white with the white inhabitants, either in church or state."

In the second half of the 19th Century the organized church began playing a larger role in the life of the Boers. But for some, like Paul Kruger who was President of the Transvaal between 1884 and 1900, the Dutch Reformed Church was not forceful enough in its rejection of English values and ideas. Ultimately Kruger and others rejected the liberalism of the Dutch Reformed Church when it permitted the singing of evangelical hymns in its worship services, a decision which the Calvinists considered to be doctrinally impure, and they established the Reformed Church, which became known by its nickname of the "Dopper" Church. It was the Doppers who encouraged the use of the term "Afrikaaner" to identify the true Boers - that is, those farmers of Dutch and German descent who were willing to follow the strict theology of Calvinism. The Doppers during this period were characterized by their aversion to all things British, and they urged true Afrikaaners to consider themselves separate and distinct as a people.

Because of the intelligence and assertiveness of their leaders, the Dopper Church became the primary exponent of this new Afrikaaner ideology. The Doppers established a theological school and wrote extensively on the application of Calvinist doctrine to their particular circumstances. In general it can be said that the Dopper Church prepared the Afrikaaner community for complete separation from the English that occurred after the Boer War.

In the two decades following the Boer War, the Afrikaaners, as they began to commonly call themselves, coalesced into a tightly defined and cohesive political and cultural people. The causes of this coalescence can probably be attributed to the fierce spirit of nationalism following the end of the war, coupled with principles of Calvinism that suggested to the Afrikaaners first that they were an elect people; and second that the Will of God directed that they separate themselves from other peoples, especially from the British, and that they maintain their racial and cultural purity. Afrikaaners were appalled by the message brought to South Africa following the Boer War by Methodist missionaries declaring that conversion to Christianity was all that was necessary in order to make a Black African the equivalent of a European.

One of the primary catalysts in the development of this new Afrikaaner spirit was the poetry of a South African theologian in the Dutch Reformed Church named J.D. duToit, known by the pen name of Totius. His poems depicted the struggle of the Boers against the Blacks and against the English, and it depicted the Blacks as particularly cruel and brutal. The poetry referred often and in glowing terms to the Boers as a chosen people, and to the Great Trek of 1835-1839 as being the second exodus from the bondage of the British. The Battle of Blood River was glorified in his poetry as an event in which God made a new covenant with His chosen people. Totius characterized the Afrikaaners as individualists with a strong sense of identity and nobility, stating in one of his poems, "The hallmark of their nobility is the purity of their race and blood."

Having distanced themselves from the English and their churches on theological grounds, the Afrikaaners then set about assuring the continuation of their separateness by pushing for the establishment of an independent language for Afrikaaners, what we know as Afrikaans. This particular mixture of Dutch, German, French and native African words did not take on the dignity of a separate language until the 1870's when a systematic effort was made to record its words and their usage. But by 1925, at the dogged insistence of the Afrikaaners that a separate language was necessary, Afrikaans was officially accepted as the second language of South Africa. Their efforts to establish a separate national identity were nearly complete.

With their own language, Afrikaaners could now insist that their children be educated primarily in Afrikaans. This, they felt, would further assure the transfer of Afrikaaner ideology and values from generation to generation in the pure and unadulterated form, especially without taint from English interference.

During the years after the Boer War Afrikaaners became more politically assertive as well. In 1918 a small group of ministers, or dominees, of the Dopper Church and the larger Dutch Reformed Church, together with other influential Afrikaaners, formed an organization known as the Broederbond. Originally founded as an Afrikaaner mutual-help association, it became a secret society in 1924. The Broederbond was and continues to be one of the most influential forces within the Afrikaaner community. Significantly, at least until recently, no less than 20% of its membership was comprised of ministers and theologians of the Dutch Reformed Church. On the broader political stage, the more assertive Afrikaaners allied themselves behind the first National Party founded in 1914 by General James Hertzog. In its program the Party defined itself as "Christian Nationalists" seeking to develop a South African national life in accordance with the major principles of the Christian religion and to attain independence in the future. Its native policy (the term "Bantu" was not yet in vogue) emphasized the trusteeship of the European population over the primitive peoples of Africa in a Christian spirit. The National Party obtained the unswerving support of the Doppers (approximately 10% of the Afrikaaner population) and over the years its support steadily grew until, as mentioned earlier, it was merged with Jan Smuts' party in 1934 in the wake of the world wide depression.

As a result of this merger, a dominee of the Dutch Reformed Church named Daniel F. Malan broke ranks and formed a new, purified National Party. It was Malan's party that in 1948 rode to victory under the banner of Apartheid. From the outset the doctrine of Apartheid as expressed by Malan had the enthusiastic approval of eminent theologians of the Dutch Reformed Church. Following Malan's victory in 1948, the Dutch Reformed Church in South Africa formally and wholeheartedly endorsed the doctrine of Apartheid, a position from which it has not retreated.

One of the first tasks of the Afrikaaner theologians was to justify the doctrine of Apartheid within the theology of the church. As you can imagine these theological efforts are nothing short of incredible, and a few examples might serve to underscore the incongruity of Apartheid with what we would understand as Christian theology.

A Dutch Reformed Church minister and theologian, G. A. Cronje, declared in 1945:

The racial policy which we as Afrikaaners should promote must be directed to the preservation of racial and cultural variety. This is because it is according to the Will of God . . . . The more consistently the policy of Apartheid could be applied, the greater would be the security for the purity of our blood and the surer our unadulterated racial survival.

Some of the efforts at justification are a bit subtler, but often tend to strike the reader as nothing more than double talk and nonsense. For example, another Afrikaaner theologian, Professor A. B. Du Preez, presents an argument that the Homelands policy of Apartheid is good because it promotes Christian humility in Black Africans. He states:

The policy of separate development so as to attain cultural independence and to be able to serve one's own people, promotes particularly the Christian virtue of humility in the educated Bantu or Coloured so that he does not raise himself above his less privileged fellows but seeks to serve them with his spiritual gifts and education and to help them achieve the highest possible culture. Whereas the policy of integration has the effect of making the educated Bantu or Coloured snobbish as they believe that their education gives them the right to ally themselves with the Europeans and to look down on the uneducated members of their own race, the policy of separate development aims at the opposite.

In another part of his book, Dr. Du Preez presents the argument that the notion of equality is not ordained by God but is of human origin and that it must yield to the higher principle of Christian justice (a decidedly Calvinistic point of view). He states:

Christian justice must be shown to all men whom God has given us as neighbors and with whom we have to live. This concept of Christian justice is much higher than the humanistic claim for equality because equal treatment for white and nonwhite usually results in the weaker race being treated unfairly and suffering injustice just as happens when equal treatment is meted out to an adult and an immature child. Our hope of survival depends upon our treating the Bantu in a just and Christian way.

These same theologians also carefully constructed elaborate and somewhat preposterous positions that sought to establish scriptural justifications of Apartheid, principally through the Old Testament stories of the Tower of Babel and Noah's curse upon Ham and Canaan (the supposed father of the Black race). All of these arguments are typical of the sophistry by which theological justifications of Apartheid are characterized. When you think about it, they really are eloquent statements to the abject failure of Apartheid as an idea and why it has been completely discredited outside of the Afrikaaner church.

Since the victory of the National Party in 1948, the Dutch Reformed Church and the government have marched in lockstep in advancing the doctrine of Apartheid. This is understandable since Malan and nearly all the members of his Cabinet were ordained ministers in the church. Moreover, the three succeeding Prime Ministers were either dominees of the church or had dominees within their immediate families. Just as the South African government has incurred the opprobrium of many of its own citizens and most of the outside world, so has the Dutch Reformed Church. It has responded to theological challenges to Apartheid by withdrawing from organizations such as the World Council of Churches that have served as a vehicle for such challenges. Within South Africa it has refused to recognize criticism of Apartheid by any other protestant denomination. And, although the church has made limited concessions to liberalize the doctrines of Petty Apartheid, it has remained resolute in its position of support for all other aspects of Apartheid. Slowly, and by degrees, however, members and ministers of the Dutch Reformed Church are finding it difficult to square their understandings of the Christian ethic with the doctrines of Apartheid. In particular, the story of one man symbolizes both the fierce resistance of the church to change and the crisis of conscience that Apartheid and the church have created for a growing number of its ministers.

In 1963 the Reverend Christiaan Frederick Beyers Naude was perhaps one of the three most important and influential members of the Dutch Reformed Church. He was at that time Moderator of the church in the Transvaal, Moderator being the hierarchical equivalent of a Catholic Bishop. Born in 1915 to an Afrikaaner hero of the Boer War and a dominee in the church, Naude's father was also a founder of the Broederbond and was considered one of the patriarchs of the Afrikaaner cause. Beyers Naude followed in his father's footsteps, joining the Broederbond at an early age and moving quickly through the church hierarchy to a position of authority. Then, at the pinnacle of his achievement, his world view began to change, primarily as a result of a tour of Europe and North America as an emissary for the Dutch Reformed Church. On that trip he realized that his own knowledge of what was happening in South Africa was full of blank spaces. Churchmen in all the countries he visited asked him about the African National Congress. He had heard of it only vaguely. Names of Black leaders were mentioned in his presence, and he realized he barely knew their names, much less what they stood for. He was asked what he thought of Alan Paton's Cry, the Beloved Country, and he confessed that he had heard of neither the author nor the book.

For Naude that trip marked the beginning of a pilgrimage that would take him from one end of the South African system to the other. As a result of his awakening, he returned with a determination to learn the true condition of Blacks in South Africa and was so appalled by what he found that he resigned as Moderator of the church. At the same time, rumblings began to occur in the church establishment because his sermons appeared to some to be too liberal, appeared to be suggesting, ever so slightly, that the whites should find a way to accommodate Blacks. With mounting hostility from the church hierarchy, Naude resigned all his formal positions with the church and also resigned from the Broederbond, and in 1963, he founded an organization called the Christian Institute, his idea being that an elite group of idealistic Afrikaaner clergymen could come together under the aegis of the Institute and gradually change the position of the church and its people.

But the church's reaction to the establishment of the Institute was immediate and brutal. Naude was unilaterally dismissed as a minister in the church and was branded as a traitor and heretic to his people. In 1966 the National Council Against Communism, under the Chairmanship of the Reverend J.D. Vorster, brother of the then Prime Minister, accused the Christian Institute of "undermining the spiritual resistance of the Afrikaaner people, working in collaboration with the Catholics, the protestants' age-old enemies, and holding multi-racial meetings strictly forbidden in Holy Scripture."

In 1971 and again in 1973, the Dutch Reformed Church brought to bear the power of the Apartheid laws against the Christian Institute. In those years the Institute, which was moving toward greater Black participation in its activities, became subject to massive searches. Naude had given financial support to Steve Biko and the program for which he worked. He had also appointed to the Institute's governing body such Black leaders as Dr. Manas Buthelezi, a noted Lutheran theologian and cousin of Gatsha Buthelezi, Chief of the Zulus, and Jane Phakati, Chairman of the South African YMCA. As a result of her activities with the Christian Institute, Jane Phakati was imprisoned during 1976 and was tortured. She managed to leave South Africa clandestinely in May, 1977.

In May, 1975, the Christian Institute was declared an "affected organization" under the Unlawful Organizations Act, thus limiting its ability to communicate with its followers and to raise funds at home or abroad. In 1976, the Institute launched a public appeal where it called upon the government to consider a national convention that was representative of all peoples of South Africa, and it reaffirmed its support for all peaceful efforts to bring change to the system of Apartheid. As a result of these efforts, the Institute was declared an "illegal organization" in October, 1977, just five weeks after Steve Biko died in prison. It became a criminal offense to refer to the Institute or to quote anything it had published. The Christian Institute was banned as an organization and Beyers Naude was banned as an individual.

As one writer states, "Naude's apostacy was especially galling and unforgive able not only because it had occurred at the very heart of the Afrikaaner power elite, but also because Naude was not content to be merely a voice of conscience. Instead, he had crossed that invisible line the separates liberals from radicals in the eyes of most Afrikaaners: he had sided with the Blacks." The spiritual and political pilgrimage of Beyers Naude still remains somewhat unique in South Africa. But it does offer some slight hope that others will follow his courageous example.

SELECTED BIBLIOGRAPHY

Cornevin, Marianne, *Apartheid: Power and Historical Falsification, United Nations Educational, Scientific and Cultural Organization,* 1980.

Du Preez, Professor Dr. A. B., *Inside the South African Crucible,* H.A.U.M. - *Kaapstad-Pretoria,* 1959.

Hexham, Irving, *The Irony of Apartheid: The Struggle for National Independence of Afrikaner Calvinism Against British Imperialism,* The Edwin Mellen Press, New York and Toronto, 1981.

Lamb, David, *The Africans,* Vintage Books, 1984.

Leith, John H., *Introduction to the Reformed Tradition,* John Knox Press, 1977.

Lelyveld, Joseph, *Move Your Shadow,* Times Books, 1985.

Southern Africa Working Party, *South Africa: Challenge and Hope, American Friends Service Committee,* 1982.

Michener, James A., *The Covenant,* Fawcett Crest, 1980.

University of California Press, *South Africa: Time Running Out, The Report of the Study Commission on U.S. Policy Toward Southern Africa,* 1981.

Ungar, Sanford, Africa, *The People and Politics of an Emerging Continent,* Simon and Schuster, 1985.

The Round Table Presentation | June 2, 1988

The Light Fantastic
Exploring The Phenomena Of Light

As the title suggests, this is a paper about light. As we sit together in this room, we are aware that light surrounds us. But except when we are struck by the reflective glare of a bright light, we normally don't think much about light. We see objects, and colors, and motion. certainly don't talk about seeing "light." Occasionally, when we see some of the peculiar aspects of the behavior of light - the distorted view of objects in water, or the left-handed image that we see in a mirror we realize the existence of something between the object and our perception of the object.

This paper traces, in the briefest way, how mankind has come to understand the behavior of light and, within the past few decades, what light really is and the role it plays in our world.

It is believed that crude lenses and mirrors have been used as far back as recorded history to concentrate the light of the sun and create fire. Etruscans and Greeks were known to use "burning glasses" for just that purpose. These consisted of shallow clear glass dishes that were filled with water.

The Greek philosophers of the golden age were less interested in the nature of light than in the mechanism of vision, and as their writings demonstrate, they were short on facts but long on speculation. Pythagoras suggested that light consists of small particles sent out by objects toward the eye of the viewer, and that turned out to be not far off the mark. Plato, however, argued that vision consisted of two actions: one emanating from an object and the other emanating from the eyes. Vision, said Plato, is the meeting of these two emanations.

It was Euclid, the mathematician, a pupil of Plato and a younger contemporary of Aristotle, who first began to express his observations about light in a disciplined fashion. In his Optics, he measured angles and properly stated the law of reflection, that is, a light ray (a term attributed to Euclid), is reflected from a plane surface at the same angle at which it strikes the surface. He studied reflections from convex and concave mirrors and also postulated that light travels in straight lines.

Galen, at the end of the Second Century A.D., made important contributions to the subject of vision and light with his anatomical investigations. Although many of his conclusions were incorrect, his determination that the back of the eye was the primary receptor and interpreter of light was an important conceptual advance.

Also in the Second Century A.D., in Alexandria, Claudius Ptolemy made the first systematic study of refraction (that is, the bending of light through substances such as glass or water). Ptolemy, of course, is known for the Ptolemaic system, in which the sun and planets orbit the earth, a system of cosmology that became accepted for the next 13 centuries. Ptolemy observed that light is bent toward the perpendicular when it enters a dense medium at an angle from one less dense and he suggested that to determine correctly the true location of a planet at a given time, one must take account of the bending of the light as it enters the earth's atmosphere.

The next major figure to inquire into the nature of light was one of the most notable scientists and philosophers ever produced by the Islamic empire, Abu Ali Al-Hasan, known in the west as Alhazen. At the end of the 10th Century, Alhazen, living in Baghdad, one of the great centers of learning in Islam, published two works in which he carried out a complete, and generally correct, mathematical analysis of convex and concave mirrors. Following the lead of Ptolemy with regard to refraction of light through the earth's atmosphere, Alhazen determined that twilight, the persistence of daylight after the sun has set, was due to the refraction of sunlight through the atmosphere. He made the assumption that twilight ends when the sun's rays are refracted from the very top of the atmosphere and then deduced that the atmosphere was 20 to 30 miles high, a good estimate even by current standards.

Fortunately for the west, Alhazen wrote in Latin and his works were available to some of the medieval scholars such as Roger Bacon and the scholars of the renaissance such as Leonardo Da Vinci.

Leonardo may have been the first scientist to utilize the camera obscura (dark room) and to hypothesize that the eye worked in a similar fashion. He wrote:

> [Let] *the image penetrate through a small round hole cut into a very dark room. You will then receive the image upon white paper placed within this dark room and . . . close to the hole, and you will see all the aforesaid objects upon the white paper with their actual forms and colors but they will appear smaller and upside down . . . . The same happens inside the pupil.*

We pass by Copernicus, the invention of the telescope in 1608 by Hans Lippershey, and its use by Galileo, and we come to the first major advance in the discovery of the nature of light, by none other than Isaac Newton. Newton's most celebrated contributions to science, of course, were his general laws of motion and his theory of universal gravitation for which he was lionized during his lifetime. But he also made important investigations into the nature of light.

Newton was born in 1642 and was only 24 years old when he began his study of colors and optics, and formulated his theory of the nature of light. In a letter to the Royal Society several years later, Newton describes how it all began:

> *[I]n the beginning of the year 1666 (at which time I applied myself to the grinding of optick glasses of other figures than spherical) I procured me a triangular glass prism, to try therewith the celebrated phenomena of colors. And in order thereto having darkened my chamber and made a small hole in my window shuts, to let in a convenient quantity of the sunlight, I placed my prism at its entrance, that it might be thereby refracted to the opposite wall. It was at first a very pleasing divertisement, to view the vivid and intense colors produced thereby; but after a while applying myself to consider them more circumspectly, I became surprised to see them in an oblong form; which according to the received law of refraction, I expected should have been circular.*

Newton's conclusion from this observation was, quite correctly, that white light is comprised of various components of colored light which, when passing through a prism, are refracted (bent) to different degrees, thus allowing a circular spot of light to be spread into an oblong spectrum of color. Using double prisms, Newton demonstrated that each color produced in the original spectrum could not be further broken down, and, in a separate experiment, he passed the colored spectrum created by a first prism through a second, inverted prism which caused the spectral light to recombine into a single spot of white light.

Newton presented in his book "Opticks" a broad-based explanation of the nature of light. Given Newton's mechanistic view of the universe, it should be no surprise that he concluded that light consisted of tiny particles or, as he called them, The various colors of light in the spectrum were to be differentiated by "corpuscles." size, with red corpuscles having the largest mass and size, and violet the smallest. According to Newton light travels in a straight line, it casts sharp shadows, and it behaves in all respects in accordance with his fundamental laws of motion.

But Newton was confronted with phenomena which his theory could not explain. For example, if light consists of particles of definable mass, when two beams of light cross one another, why don't the collisions of the particles cancel or disrupt or dim the light beams? If light travels in a straight line, why is it that experiments show a fuzzy area along the edge of a shadow? And if light consists of particles, how can Newton's particle theory explain the phenomenon that exists when light from a single source passes through two small holes into a darkroom where the separate light beams spread out and partially overlap with each other on a reflective screen? Under Newton's particle theory, the overlapping area should be much brighter since twice as many particles are striking that area as are striking the two independent areas. But in fact, experiments showed that rather than being brighter, this common area was covered with alternating light and dark lines.

Newton could not explain these phenomena and he somewhat glibly dismissed them, but a contemporary of Newton had an explanation. Christian Huygens of The Hague in the Netherlands, made his life's work the study of light. He postulated that light, rather than particles, consisted of waves, much like waves that emanate from dropping an object into a pool of water. Huygens demonstrated that waves can cross each other going in different directions and not lose momentum, regularity or direction. Moreover, waves undergo diffraction, that is, they bend around corners, thus explaining the fuzzy areas along the edges of shadows. Most significantly, separate waves when they encounter one another, and especially when they are traveling in a similar direction, can, during the period they are traveling together, either reinforce or neutralize one another. That is, the crest of one wave can converge with and reinforce the crest of the other wave, and we get a more pronounced crest, or the crest of one wave can move together with the trough of the second wave, thus neutralizing the effect of both waves and leaving the water level undisturbed. This is called interference: destructive interference if two waves cancel, and constructive interference if two waves add. The series of light and dark lines that appeared when light from a source passed through small holes and was allowed to converge upon a screen could clearly be explained by the law of interference.

Nonetheless, given Newton's reputation, Huygens' theory never had a chance. Newton simply could not be wrong. After all, he had explained the fundamental forces of the universe and established a set of simple mechanical laws that so wonderfully described not only the physical world but the motions of the planets as well. Moreover, Huygens' theory was not nearly as well developed mathematically as was Newton's. Also, Huygens' theory itself encountered a formidable obstacle: if light consisted of waves, then how could light travel through a vacuum? What could carry the waves? Huygens' answer was ingenious, but not very satisfactory. He resurrected the Aristotelian concept of ether, a substance that Aristotle claimed was the fifth essence, the "quintessence," a substance that filled the universe, was invisible, perfectly stationary, perfectly permeable and absolutely undetectable. Then, said Huygens, light could travel in waves across a vacuum because the vacuum was in fact filled with ether and the ether could carry the waves. As an aside, one of the great challenges for the 19th century physicists was to confirm the presence of ether. All efforts ended in failure and finally Einstein established that whether or not ether existed didn't really matter because the physical laws of the universe could be fully explained without it.

It was during the time of Newton and Huygens that one of the most significant discoveries with respect to the behavior of light was made - its velocity. Throughout ancient and medieval times, the velocity of light was assumed to be infinite. Galileo was the first we know to question this, and in a very crude way he set about to measure the speed of light. He placed two people on hilltops approximately a mile apart, both carrying shielded lanterns. One was to uncover his lantern and the other, upon seeing the light, was at once to uncover his own lantern. If the first man measured the time that elapsed between his own uncovering and the sight of light from the other hill, he would know how long it took light to make the round trip. It is easy to see why that experiment was a total failure.

But in 1676 the first realistic measurement of the velocity of light was made by a Danish astronomer, Olaus Roemer, who was working at the Paris observatory observing the satellites of Jupiter. Their times of revolution had already been carefully measured, so Roemer thought it possible to predict the exact moments at which each would pass into eclipse behind Jupiter. Roemer conducted these measurements at numerous times throughout the year and, to his surprise, the moons were being eclipsed at different times. At those times of the year when the earth was approaching Jupiter, the eclipses came earlier and earlier, while when earth was receding from Jupiter, just the opposite was observed. Roemer reasoned that he did not see the eclipse when it took place, but only when the cut-off end of the .light beam reached him. The orbit of each moon had not changed, and the eclipse itself took place at the scheduled moment, but when the earth was closer to Jupiter Roemer saw the eclipse sooner than when the earth had moved further away from Jupiter. The difference in time between the earliest and latest eclipse must therefore represent the time it took light to travel the diameter of earth's orbit. Roemer measured the time differential and using the best figure he knew for the diameter of the earth's orbit, calculated that light traveled at a velocity of 138,000 miles per second, approximately 3/4 of what is now accepted as the correct value.

It was approximately 200 years later when the first accurate measurement of the velocity of light on earth was made. This was done by a French physicist, Armand Fizeau, who went back to Galileo's method, but eliminated the personal element. In 1849 Fizeau set up a large, rotating wheel on one hilltop and placed a large mirror on another hill five miles away. Fizeau had cut 740 teeth into the edge of the wheel so that it resembled a giant gear. He placed a light behind the edge of the wheel facing toward the mirror five miles away. The light was small enough to be blocked from shining on the mirror by each tooth on the wheel. But at each gap in the wheel, the light passed through and was reflected by the mirror. If the wheel turned at the proper speed, the reflected light pulse from the mirror returned just as the next tooth moved into line, and the light was blocked. The result was that, to a person sitting stationary behind the wheel, no light would be seen reflected from the mirror. From the velocity at which the wheel had to be turned in order for the returning light pulse to be blocked, it was possible to calculate the time it took light to cover the ten mile round trip distance. Fizeau was five percent off in his calculation, but later experiments of a similar nature have pinpointed the velocity of light just short of 300,000 kilometers, or 186,200 miles, per second.

For nearly 150 years, Newton's particle theory of light held sway until it was challenged and finally disposed of by another Englishman, Thomas Young - who has to be one of the most interesting characters in the history of science. Thomas Young is considered by some to be the last man who knew everything. Born in 1773, he was an infant prodigy. He could read at the age of two and had worked his way twice through the Bible at four. Before he was 15 he had learned a dozen languages, including Greek, Latin, Hebrew, Arabic, Persian, Turkish and Ethiopian. He could also play more than a dozen musical instruments, including the bagpipes.

Known as "Phenomenon Young" at Cambridge, he took up medicine, but was generally unsuccessful because he apparently could not get along with anybody, including his patients. As a man in his 20's, he pursued researches in physics and mathematics and occupied the Chair of Natural Philosophy at the Royal Institution. He also contributed to the fields of botany, philosophy, language, and physiology, especially with respect to the mechanism of vision. He developed what is known as the trichromatic nature of color vision, proposing that the retina has three different types of color receptors, each with a different wavelength sensitivity - in red, blue and yellow. And as if that were not enough, in 1814 he traveled to Cairo and made one of the most significant contributions to archeology by translating Egyptian hieroglyphics by using the Rosetta stone as a key.

In 1801 at the age of 28, Young presented to the Royal Society of London a lecture entitled "The Theory of Light and Colors." He came down squarely in support of the wave theory of light and supplied much of the mathematical analysis that Huygens had lacked. He showed that light waves behaved like water waves – transverse waves - rather than sound waves which are pressure waves. Colors, said Young, represented different wavelengths (or frequencies) of light, blue being the shortest wavelength (highest frequency) and red the longest (lowest frequency). Young presented a method of calculating the different wavelengths of light and more than a century later using much more sophisticated equipment those calculations were proven to be accurate to one-thousandth of a millimeter. In this single presentation at age 28, Young penetrated many of the secrets of light, and he succeeded in elevating the wave theory of light into predominance.

During the 19th century, following Young's brilliant contributions, discoveries regarding the nature of light came rapidly. First, in the early part of the century, the Danish physicist, Hans Oersted, through his investigations into electricity and magnetism, established that the two forces were intimately related. This actually occurred quite by accident during the course of a lecture by Oersted when he happened to place a current carrying wire over a compass and saw that the needle violently reoriented itself in a manner perpendicular to the wire. Within only a few years, thanks to some innovative work by Michael Faraday, scientists were talking about electromagnetism as a unified force. But the results of these experiments with electricity and magnetism failed for a number of years to receive any theoretical justification. Faraday, for instance, perhaps the greatest electrical innovator of all, was completely unskilled in mathematics, and was unable to develop a mathematical basis for this new found electromagnetism.

Into the void stepped James Clerk Maxwell, a Scotsman who was a mathematical genius, and who, at the age of 15, contributed a piece of original work to the Royal Society of Edinburgh that was of such quality that the members of the Society refused to believe that the boy was the author.

In the 1860's, after working out the kinetic theory of gases, a major scientific achievement in itself, Maxwell worked out a set of four comparatively simple equations that have been known ever since at "Maxwell's Equations." These equations established in a simple and direct manner that it was impossible to consider an electric field or a magnetic field in isolation. The two were always present together, directed at mutual right angles, so that there was a single electromagnetic field. These equations allowed Maxwell to calculate the velocity at which a field electromagnetic wave would have to move. This turned out to have a value of 300,000 kilometers per second, precisely equal to the velocity of light. Maxwell knew this wasn't just a coincidence, and he concluded, quite correctly, that light itself was an electromagnetic radiation. The equations pointed out that all electromagnetic radiation traveled at the same velocity but had different energy levels (i.e., frequencies) depending on the comparable oscillations of the source of the energy.

In other words, the frequency of light established how much energy it had. Incidentally, Maxwell is considered to be primarily responsible for the invention of television since his equations formed the theoretical foundation for its later development.

In 1888, at age 29, the German physicist Heinrich Hertz discovered radiation that we now call radio electromagnetic radiation of very low frequency waves. This discovery completely bore put Maxwell's prediction and was accepted as evidence for the validity of Maxwell's theoretical equations. It was ironic that Hertz did not believe his "electric" waves would have any practical application. In the early 1890's an Italian inventor named Marconi read about Hertz's work and imagined communication by Hertzian waves. Marconi's first wireless message was sent in 1895, just a year after Hertz's premature death. In 1895, another German physicist, Wilhelm Roentgen, discovered electromagnetic radiation of very high frequency - radiation that we now call X-rays. And in 1898, Marie Curie discovered ultra high frequency gamma rays emitted as a result of radioactive decay.

So, let's see where we are at the end of the 19th century. The wave theory of light has gained complete acceptance and the particle theory has been discredited. Maxwell's equations have confirmed the wave theory and have unified the forces of light, electricity and magnetism, and the equations have been born out in experiment after experiment. What's more, visible light was found to comprise less than two percent of the entire electromagnetic spectrum which, beginning with the longest wave lengths, consists of the following:

> *Radio waves: frequencies on the average of 1,000,000 cycles per second, wavelengths average 300 meters. Short wave radio uses frequencies 10 times as great (1 billion cycles per second) and TV 10 times greater than short wave radio.*

Microwaves: frequencies from 1 billion to 100 billion cycles per second (wavelength from 30 centimeters to .3 centimeters. Radio telescopes and radar use this range.

Infrared rays: frequencies from 100 billion to nearly 1 quadrillion.

Visible light rays: a small band of frequencies just under 1 quadrillion.

Ultraviolet rays, x-rays, and gamma rays: frequencies go up to 100 quintillion cycles per second and wave lengths of 1 billionth of a centimeter.

The beginning of the 20th century marked a decisive turning point for physics. The quantum theory was born and scientists began to unravel the mysteries of the atom - how it could be described, how it behaved, and how that behavior could be related to the physical world around us. While the quantum theory and atomic physics are beyond the scope of this paper, at the heart of this new physics is the behavior of light. That is to say that light plays a fundamental part in all atomic interactions, because it serves to transfer energy from one atom to another. And it should come as no surprise that the person primarily responsible for unlocking the mysteries of this new physics and describing how light plays a role in it was Albert Einstein.

In 1905, Albert Einstein published three papers in a Berlin physics journal. Einstein had not yet received his doctorate and would not until 1906. Hence, when these papers were published, Einstein had no affiliation with any university or educational institution. He was an unknown.

Any of the three papers that Einstein published in 1905 would have assured him a place in the annals of science. The first paper dealt with the phenomenon of "Brownian motion," (droplets of water containing pollen grains that were seen to bounce around when viewed under a microscope). Here Einstein furnished the mathematical and theoretical underpinnings for the proof of the existence of atoms. The second paper presented Einstein's theory of special relativity, and established the speed of light as the maximum velocity for anything in the universe. And through the formula $E=mc^2$ he demonstrated that the conversion of mass to energy or energy to mass was inextricably bound up with the emission or absorption of light.

Einstein's third paper was on something called the photoelectric effect. This was Einstein's attempt to explain a phenomenon that scientists could not account for. When light was focused on metal surfaces, electrons were found to be ejected from those surfaces, as if the light rays literally knocked them out of their atoms and they became free electrons. Einstein's paper on the photoelectric effect provided science with its first description as to how light played a part in the transfer of energy at the atomic level. It is interesting to note that while Einstein will forever be associated first and foremost with his theory of relativity, when he received the Nobel Prize in 1921, it was for his paper on the photoelectric effect and his contribution to quantum physics.

As you can imagine, this stuff is far too complicated to try and explain in just a few minutes. But at the risk of causing Einstein to turn in his grave, let me give you a brief description of Einstein's revolutionary view of matter and energy on the atomic scale. Einstein suggested that atoms throughout the universe exchange energy by absorbing and emitting pulses of light. He called these pulses of light photons and photons represent the smallest particle that can possibly exist. Atoms that are full of energy emit photons which instantaneously begin travelling at 300,000 kilometers per second in straight lines until deflected by other particles or until they are absorbed by other atoms in their paths ready to take on more energy. Einstein also said that while all photons travel at the speed of light, photons can differ in the amount of energy they possess. Some photons would have a low frequency, such as infrared light, and others would be of higher frequency such as ultraviolet light. Einstein also theorized - and this turned out to be crucial - that the atoms of each element only emitted and absorbed light of particular frequencies. Sodium, for example, only releases or absorbs yellow light, neon only red and orange light, copper only green and blue light. Einstein's theories of light had immediate acceptance because they explained so many unanswered questions. Now it became clear why photographic film could be ruined if exposed to normal light, but not if the light were only low energy red light. The frequency, or energy, of red light was not enough to cause the necessary chemical reactions in the film. And photosynthesis could now be explained. Prior to Einstein's paper people did not know why sunlight caused plants to grow, but when plants were placed indoors in the presence of high intensity incandescent light, the plants withered and died. Ultraviolet light furnished by the sun is necessary to cause the chemical reactions that are part of the process of photosynthesis. Lower energy incandescent light can't do the job. And Einstein finally provided a theoretical foundation for the science of spectroscopy. In the 19th century scientists knew that different metals, when heated to incandescence, emitted only a narrow frequency of visible light, sodium in the yellow, neon in the red and orange, and so forth. Einstein's theory explained this completely. In fact, the picture Einstein created of what happens at the atomic level was so simple and so elegant that it gave immediate credibility to the quantum theory that had been espoused by Max Planck a few years before and turned the attention of most of the world's leading physicists to the field of what was to become known as quantum physics.

As I mentioned, central to his theory is the existence of something called a photon. But what is a photon? According to Einstein a photon must be a particle of light - or electromagnetic radiation. It must be a particle and not a wave, although Einstein admitted that it may have some wave-like properties. Photons are spontaneously emitted and absorbed by atoms and normally exist for periods of time that are unimaginably small. According to Einstein a photon is a particle of pure energy.

Now, this is the one part of Einstein's new picture of atomic interactions that the scientific community found it difficult to accept - his conclusion that light was a particle and not a wave. After all, the wave theory had been confirmed again and again during the 19th century. But Einstein was proved to be absolutely correct, and the story of how that was accomplished is of special interest to this group, for it was done by none other than John Compton's father, Arthur Compton.

In the early 1920's Arthur Compton conducted a series of experiments where he shot x-ray photons into a stream of electrons. When a photon struck an electron he was able to record and measure what happened when they collided. As it turned out, the collisions were like the impact of two billiard balls. Compton observed that both the photon and the electron were deflected, and went off at angles, and that there was a transfer of momentum and energy consistent with the collision of particles. These experiments established beyond a reasonable doubt that a photon of light had to be regarded as a particle and not a wave, and for his discovery of the Compton effect, he received a Nobel Prize in 1927.

Although Einstein insisted that the photon must be viewed as a particle in order to explain both his theory of relativity and the photoelectric effect, he was the first to admit that since quantum physics is beyond the common experience of all of us, there is no reason to suppose why a photon could not be both a wave and a particle. Rather than viewing wave and particle properties as mutually exclusive, as had been the case in classical physics, Einstein suggested that the two properties be viewed as complimentary. The work of some of the leading scientists in quantum physics during the past 30 years have reinforced Einstein's belief that the behavior of light at the quantum level cannot be fully explained without being viewed as having both particle and wave properties.

Let me add that Einstein's contributions to the science of light did not stop in 1905. In 1917 he published a paper where he suggested that emissions of light photons from atoms of the same element could be stimulated by flooding the atoms with photons of similar wave lengths, and he theorized that the result could be a beam of pure, monochromatic (a single color), highly intense light that was all in phase (same frequency and in unison). This paper served as the theoretical blue print for the laser which is an acronym for "light amplification by stimulated emission of radiation." Today the laser has become one of the most useful tools of science medicine and business. Laser surgery is becoming more commonplace. Laser weapons are part of the "Star Wars" program. And lasers are essential to communication by optic fibers - a form of communication that will increase our ability to transmit data millions of times more effectively than current methods.

One final note. Since Einstein's contributions the world of particle physics has become much more complicated. Now it is populated by such things as positrons, neutrinos, pions, mesons, and the like. But so far no one has been able to controvert the role of light - of light photons - in the fundamental transfer of matter to energy and back to matter. But it remains to be seen whether Einstein's picture represents a definitive view of light or whether it is only an intermediate, and maybe even a primitive, step in our ultimate understanding of the subject.

BIBLIOGRAPHY

Isaac Asimov, *Asimov on Physics* (1976).

Isaac Asimov, *Understanding Physics: Light. Magnetism and Electricity* (1966).

John Gribbin, *In Search of Schrodinger's Cat - Quantum Physics and Reality* (1984).

George Gamow, *Thirty Years that Shook Physics* (1966).

Vasco Ronchi, *The Nature of Light* (1970).

Michael I. Sobel, Light (1987).

The Round Table Presentation | December 5, 1991

Barbarians At The Gates
The Battle Of Thermopylae

In the style to which presentations to The Round Table have been accustomed, I also decided to give the title of this paper a double meaning. As you know, Barbarians at the Gate is the title of a recent book that tells the story of the leveraged takeover of RJR Nabisco, a story that has become the paradigm for the hubris and money lust of the 1980's.

My title, Barbarians at the Gates, and the story it tells, is also a story about an attempted takeover but not of the modern variety. The time in question is August 480 B.C. and Xerxes, the king of the Persian Empire, the greatest empire the world had then ever known, had just given the order for the invasion of Europe. Barbarians at the Gates refers to the Persian army confronting the men of Sparta at the Battle of Thermopylae. The Persians are described by Herodotus, and all the Greeks of that time, as "barbaroi" (barbarians) because they seemed to babble "bar, bar, bar" and could not speak Greek. Thermopylae, translated from the Greek, means "the hot gates" because of the hot sulphur springs that are found there.

The failure of the Persian invasion in 480 B.C. was determined by four decisive battles: Thermopylae and Artemisium in August 480, Salamis in September 480, and Plataea in August

479. Artemisium and Salamis were the two decisive naval battles, Artemisium being a draw, but an important draw, and Salamis being an unquestioned Greek victory. Plataea, fought the following summer, was so complete a Greek victory that the remnants of the Persian forces beat a hasty exit through northern Greece and across the Hellespont.

However, the most well-remembered battle of the Greco-Persian War was not a victory for the Greeks, at least not a military victory. Almost immediately after it happened, the defense of the pass at Thermopylae by Leonidas and his 300 Spartans was recognized as a glorious and heroic stand that served to unite the previously divided Greek city-states, and to bring to the Greeks their first sense of nationhood, of Greekness.

There are several important consequences of Thermopylae to which I will refer later. Most enduring may be that the stand of Leonidas at Thermopylae has come to symbolize some of the noblest attributes of the human character. Montaigne, the Sixteenth Century French essayist, wrote of Thermopylae in the following terms:

> *The worth and value of a man is in his heart and in his will; there lies his real honor. Valour is the strength, not of legs and arms, but of heart and soul; it consists not in the worth of our horse or our weapons, but in our own. He who falls obstinate in his courage, if he has fallen, he fights on his knees (from Seneca). He who relaxes none of his assurance, no matter how great the danger of imminent death; who, giving up his soul, still looks firmly and scornfully at his enemy - - he is beaten not by us, but by fortune; he is killed, not conquered.*
>
> *The most valiant are sometimes the most unfortunate. Thus, there are triumphant defeats that rival victories. Nor did those sister victories, the fairest that the sun ever set eyes on - Salamis and ... Plataea...- ever dare*

> *match all their combined glory against the glory of the annihilation of King Leonidas and his men at the pass of Thermopylae.*

The invasion of Europe by Xerxes in 480 was the culmination of 70 years of constant expansion of the Persian Empire. From a long range historical perspective the invasion (and the invasion of Darius 10 years before) can be viewed as the first in a series of struggles between East and West, between Asian and European powers that found its sequels in the empire established by Alexander the Great, and later by Rome, the sweep of Islam across the Mediterranean, the Crusades, the western incursion of the Ottoman Turks, and even the Arab-Israeli conflict can be viewed in these terms.

Established in 550 B.C. by Cyrus the Elder, like most great empires, Persia was founded upon the ruins of the Medes, Assyrians and Babylonians. By 480 B.C. its borders ran from the Hindu Kush in the east, to the Aral Sea and the Caspian Sea in the north, to the Arabian Peninsula and Egypt in the south and southwest, to Palestine and Israel, and Anatolia, or Turkey, in the west.

As part of the Persian conquest of Turkey, Cyrus defeated King Croesus in 547, and this brought the Ionian Greek cities into the Empire. These were the cities along the west coast of Turkey and included Halicarnassus, Miletus and Ephesus.

The triggering event of the Greco-Persian wars was a revolt against Persian rule by the Ionian Greek cities in 500 B.C. Within four years, the Persian army had again subjugated the entire area and Darius, then the King, selected Mardonius, his nephew and son-in-law, to secure further territories on the other side of the Hellespont. In a joint land and sea effort, Mardonius crossed the Hellespont with a relatively small force, secured Thrace and accepted the submission of Macedonia to the Persian empire. However, his fleet ran into a violent gale near Mt. Athos, a disaster that the Persians did not forget. Mardonius returned to Asia having secured the entire northern Aegean for the Persian empire, but he had failed to subdue the principal Greek city-states.

So Darius was not satisfied. The Ionian Greeks had been supported in their revolt against the Persian empire by their cousins across the Aegean, primarily Athens and Sparta, and Darius was bent on a punitive expedition in order to discourage any further revolt.

Darius elected to sail directly across the Aegean rather than make a march around its northern boundary, as Mardonius had done, and in 490 B.C. his forces sailed from Ionia for the mainland of Greece, subduing many of the Cyclades Islands as they passed. Then, having overrun the city-state of Eretria at the south end of the long island of Euboea, the fleet crossed the narrow channel and landed on the beach before the plain of Marathon. Athens, certain of the Persians' intentions when it learned they had overrun Eretria, had sent an urgent plea to Sparta for aid but were told that the Spartans could not march at once since it was the feast of the Carneian Apollo, the most sacred religious festival of the Spartans.

Repulse of the Persians was therefore left to the Athenians. Their general, Miltiades, determined rather than to fight behind the wooden walls of Athens, to march to Marathon and confront the Persian force immediately.

The Battle of Marathon, which took place in August 490 B.C., was an absolute rout of the Persians by the Athenian army. Approximately 10,000 Athenians and their allies literally pushed an army of Persians four to five times their size into the sea. Some say the Persian soldiers, having spent long hours at sea, were not physically prepared for battle so soon after landing. If Herodotus is to be believed, and he is known for exaggeration, at the end of the day less than 200 Athenians had fallen, but the Persian dead, which were carefully counted, numbered 6,400. Among those who took part in the battle of Marathon was Aeschylus, one of the greatest of the Greek poets and dramatists.

Four years after Marathon, Darius died and was succeeded by Xerxes who immediately committed to assemble the largest land and naval force that the Persians could muster to invade Europe and subdue Greece.

Almost immediately Xerxes began his preparations, and the preparations and the carrying out of this massive expedition of the Persians into Europe must have been well-known to the Greeks for years. Spies and informants were employed by both sides, although the Persians were so open with their intentions that any Greek traveler to the Ionian cities would have quickly perceived what was coming. In the several years prior to 480 B.C., in all the ports of Phoenicia, Ionia, Egypt and Cyprus, ships were being built and outfitted. The Persians engineered and dug a canal over the two mile isthmus behind Mt. Athos so as to avoid the potential destruction of the fleet that had happened to Mardonius several years before. Storage depots and granaries were constructed throughout the Persian vassal states of northern Greece in anticipation of the coming of the massive army.

Well, if the Greeks must have known what was coming, were they preparing themselves? The answer is yes, due largely to one individual. During the 10 years between the Battle of Marathon and the invasion by Xerxes, Miltiades died and was succeeded in his leadership of the Athenian assembly by Themistocles. Themistocles is considered by most historians as the greatest Greek politician who ever lived, and he was the man who was destined to save Greece from the Persian invasion under Xerxes. Thucydides, the historian of the Peloponnesian Wars, described Themistocles as follows:

Themistocles was a man who most clearly represented the phenomenon of natural genius. By sheer personal intelligence, without either previous study or special briefing, he showed both the best grasp of an emergency situation at the shortest notice, and the most far-reaching appreciation of probable further developments. He was good at explaining what he had at hand; and even of things outside his previous experience he did not fail to form a shrewd judgment. No man so well foresaw the advantages and disadvantages of a course in the still uncertain future. In short, by natural power and speed in reflection, he was the best of all men at determining promptly what had to be done.

This description is amply demonstrated by Themistocles' reaction to the discovery in 483 B.C. of a rich vein of silver near Cape Sunium perhaps thirty miles from Athens. The mines were owned by the Athenian city-state and under normal conditions, the profits would be shared among all the Athenian citizens. Themistocles, however, was able to convince the assembly at Athens that this silver should not be distributed to the citizenry but should be used to construct 100 new triremes - a new type of three-banked warship which would for some years give the Greeks and the Athenians command of most of the eastern Mediterranean. And without the triremes, the Greek victory over the Persians at Salamis could never have happened.

As I suppose all of us learned in grade school, the city-state of Sparta stood in marked contrast to that of Athens. Whereas the democracy in Athens could be characterized as demagogic, unpredictable, sometimes brutal (an example would be the custom of ostracism), Sparta, or Lacedaemon, represented a very rigid society, less individualistic than Athens and committed to a strict class structure with the highest class, the Spartiates, being dedicated to a military life from an early age. The middle class was composed of free men who fought and marched with the Spartiates but had no voting rights, and the third stratum of Spartan society was formed by the Helots, who were probably descendants of the original inhabitants and who worked on farms belonging to the Spartiates. They were not slaves but would more probably be regarded by us as serfs.

We probably all remember reading about the Spartans' commitment to the creation of a warrior race which required that sickly infants be left to die and that boys be taken at seven or eight years of age and placed in military systems where their whole life was from that point on devoted to the city-state. They were reared and educated with a focus on physical strength and fitness, discipline and war. Sparta was ruled by two kings - reminiscent of the two consuls that governed the Roman Republic. This duality was intended to be a check on autocracy. A Spartan army abroad was always commanded by one of the kings who had absolute power over his troops.

Aristotle describes the kingship at Sparta as a kind of unlimited and perpetual generalship. There was something akin to a senate or assembly but it could not debate and had little power. The society was governed by laws established by a politician named Lycurgus who had lived in the Ninth Century B.C. The Spartans observed strict religious celebrations, nearly as important as their military discipline. In fact, it was the Carneian Festival that caused them to be late for the Battle of Marathon and caused them to send only a handful of men to the pass at Thermopylae. Aristotle further remarks that the Spartan discipline produced not only superlative warriors but citizens with excellent manners. Spartans revered their elders, respected women and were taught to be modest and courteous in public. It has also been remarked that the common education and focus on discipline and military training produced a genuine equality among the Spartiates, where merit and not title or money was rewarded.

In the fall and winter of 481 and 480 B.C., as Xerxes assembled his forces on the Asian side of the Hellespont, his army undertook such an ambitious and extraordinary engineering feat that nothing before and few things since could rival it. This was the bridging of the Dardanelles, a distance of nearly two miles. Two parallel bridges were in fact created built upon nearly 700 ships serving as pontoons. The carriageway over the ships was wide enough to accommodate wagons dawn by mules and oxen and was made of timbers laid end to end covered by a foot or more of packed earth. Railings were built along its entire length to keep the animals from falling off. After several false starts and the destruction of many ships by the gales blowing down from the Russian steppes, special anchors were devised to hold the boats in place, and these bridges survived even the most violent weather of the Aegean and Black Seas until the Persians recrossed it in the following year.

Completed in May of 480 B.C., the army took many days to cross the bridge. According to Herodotus the invasion force involved 46 nations under 30 Persian generals. Warriors came from as far away as the Aral Sea, India and Ethiopia. Herodotus estimated the size of the army to be as many as five million persons, three million of whom were soldiers, but it is clear that this number is grossly exaggerated. Subsequent historians have tried to estimate the size of the invasion force based on information from ancient Persian records as well as the ability of such a force to live off the land as it came through Thessaly, Macedonia and into Attica. The best estimates are that the army's numbers were about 250,000 fighting men with as many as 50,000 others along for logistical support and as many as 75,000 camels, mules and horses.

The Persian strategy involved moving in a coordinated manner both by land and sea. Accompanying the Persian army and staying with it along the coast of Greece were more than 1,000 ships from all over the empire — Phoenicia, Egypt, Cypress, Asia-minor and the Black Sea. More than 650 of the these ships were front line war vessels equipped for battle. The coordination of the army and navy's movements were necessary so that the navy could keep the army adequately provisioned and so that the Greek naval threat could be obliterated thus preventing the Greek navy from launching attacks behind the advancing line of the Persian army.

While the invasion was proceeding west and south through Thrace, Macedonia, and Thessaly, a league of the Greek city-states meeting in Corinth, called the Isthmian League, met in June and July to determine the strategy it should adopt. The Peloponnesians argued that all the Greeks should fall back behind the so-called "Isthmian line" at Corinth where a wall across the isthmus was being built to check the Persian advance into the Peloponnese. However, it was Themistocles who persuaded the League to abandon this strategy and instead to fight as far forward in Greece as possible. In July 480 B.C. a decision was finally reached to mount a land defense at Thermopylae. At the same time, the Greek navy, led by the Athenians, would move to the port of Artemisium on the north coast of the island of Euboea and mount a coordinated defense against the Persian navy.

At the end of July, as news reached representatives of the Isthmian League that the Persian army had passed Mt. Olympus and the Athenians were preparing to embark for Artemisium, the Spartans informed the League that the Cameian Festival, that same religious festival that caused the Spartans to be absent from the Battle of Marathon, would be observed in August and therefore the entire Spartan army could not march until the festival was over. After pleas were made that Sparta send out some of its army to defend the pass at Thermopylae, Leonidas, one of Sparta's two kings, determined that he would take a small force and march to the pass, collecting what other contingents he could along the way. Leonidas, then a man in his early fifties, hand picked a force of 300 Spartiates and they were accompanied by approximately 1,000 Helots, the peasant class. As his force marched northward to the pass at Thermopylae, Leonidas gathered approximately 5,000 additional soldiers from the various city-states through which he passed.

Leonidas and his force of less than 7,000 fighting men reached the pass at Thermopylae on August 12, 480. It is worthwhile describing what they saw when they arrived. The pass is located on the southern coast of the Malian Gulf, where an ancient road running east-west snaked along the edge of a cliff some fifty feet above the sea for a distance of two miles. Herodotus described the coastline at the time as being ironbound and inhospitable. There was no beach, just a sheer drop on their right from the small road fifty feet into the sea. To their left along the two mile stretch rose the sheer cliffs of Mt. Kallidromos. The space between the cliff wall of Kallidromos and the drop-off to the sea is never more than a few hundred yards wide over this entire stretch, but at three points, known as the west, middle and east gates, the width can be counted in feet. The narrowest defile was the west gate, but there the sides of Mt. Kallidromos were not so sheer, and Leonidas feared his flank could be turned by the Persian army scrambling over the steep but not impassible hill to their left. Accordingly, Leonidas chose to make his stand at the middle gate where the pass was no more than 30-50 yards across and the wall of Kallidromos rose vertically.

At the middle gate stood an ancient wall called the Phocian Wall, that had been an ancient defensive barrier at the middle gate. Leonidas set about strengthening this wall. His forces made camp just beyond the east gate, perhaps one-half mile distant. It should be noted that Thermopylae looks quite different today than it did nearly twenty-five hundred years ago, and the same may be said for much of the rest of Greece. Mt. Kallidromos, as with most of the mountains throughout Greece, was then covered with hardwood forests. Now it and much of Greece are bare of trees, due in great part to the deliberate deforestation over the centuries for home building and ship building and similar uses. This deforestation has also caused enormous soil erosion and such extensive siltation into the Aegean that today the sea is perhaps three miles distant from the cliff's edge at Thermopylae.

Having chosen where to make his stand, Leonidas then conducted several raids into the area immediately to the west and north of the pass, pillaging and destroying fields and granaries not only to supply himself, but to leave as little for the Persians as possible. By August 14, advance guards of Xerxes' army reached the Malian plain within a mile or two of the pass, and over the next several days the entire army arrived and made camp. Herodotus tells us that Xerxes reconnoitered the pass and took a proposal to the defenders assuring them that if they laid down their arms and became allies of the Persians they would be rewarded with far richer lands that those they now possessed.

A number of Greeks, on seeing the immensity of the Persian army, argued that the subjugation of all of Greece was inevitable, that no force could be assembled to withstand the sheer size of the force Xerxes commanded, and that surrender on such generous terms was the only alternative. Leonidas rejected the offer out of hand. Herodotus then goes on to tell a story about one of the Persians reconnoitering the line of defense:

> *Xerxes sent a horseman to find out the strength of the Greek force . . . . The Persian approached the camp and made a survey of all that he could see and to observe what the soldiers were doing. This was not, of course, all of the Greek force, for he could not make out the troops behind the reconstructed and guarded wall. Nevertheless he took careful note of those troops who were stationed on the west side of the wall. At that time they happened to be Spartans, some of whom were stripped for exercise while others were combing their hair. He watched them with astonishment and took due note of their numbers and then rode back at leisure. No*

> *one attempted to pursue him, and indeed, no one took the slightest notice of him.*

Herodotus suggests this deliberate nonchalance on the part of the Spartan soldiers is indicative of their utter fearlessness and ability to intimidate the enemy. On hearing the report, Xerxes called for the traitor, Demaratus, a former king of Sparta who lost his kingship several years before as a result of charges that he was illegitimate. Leaving Sparta, Demaratus had gone over to the Persian side and was a principal advisor of Xerxes during the invasion. When Xerxes asked Demaratus the meaning of the Spartans' conduct, according to Herodotus he answered:

> *These men are making ready for the coming battle, and they are determined to contest our entrance to the pass. It is normal behavior for the Spartans to groom their hair carefully before they prepare themselves to face death. I can reassure you on one point: if these men can be defeated and the others of them who are still at home, then there is no one else in the whole world who will dare to lift a hand, or stand against you.*

After four days of preparation, on August 18, Xerxes gave orders to begin the assault on the Spartan position. Apparently Xerxes gave no thought at the time to trying to circumvent the position or to go over Mt. Kallidromos. All indications are that he wanted his men to confront this small force directly and to annihilate them.

The first wave of attackers were Medes. They wore dome shaped helmets of hammered bronze or iron, jackets of fish scale mail and carried light wicker shields. Their arms consisted of bows with a quiver full of short arrows, short spears, and a dagger for close fighting. As many as 20,000 Medes were involved in the first charge against the Spartan position, and Herodotus writes that as they came upon the bristling barrier of spears and shields they threw themselves against it and fell and died in enormous numbers without failing to move the Spartan line.

The armor and fighting weapons of the Medes were typical of Persian arms during that period in history. And it seems incontestable that the Greeks had superiority in terms of armament. The Greek fighting soldier, known as a hoplite (the term is taken from the large round Greek shield known as a hoplon) was heavily armored. The helmet was made out of hammered bronze or iron and covered the entire head. The principal parts of the body were also covered with metal armor - the shoulders, the trunk and the legs above the knee. The shield was made of solid wood and covered with bronze or iron. The hoplite fought with two weapons, a wooden shafted spear about six feet long with a bronze or iron head. It was not a javelin to be hurled but was similar to the Swiss pike that was so feared during the years when the Swiss hired themselves out as mercenaries to most of the nations of Europe. Finally, the Greeks carried a long iron sword, often two-edged.

The Spartan mode of fighting and the technique used at Thermopylae, was to stand in a close, almost unbroken wall of armor, the shield being held on the left arm and each man protecting the right side of his neighbor. The right flank was, therefore, the weak point in the Spartan line, so the best troops were always put in this position of trust and honor. Thermopylae, however, was ideal for a Spartan hoplite battle, because the weak side was guarded by a sheer cliff to the sea.

At Thermopylae, the tactic of Leonidas was to stand firm in the opening phase and let the Persians simply break themselves against the weight and solidity of the phalanx. When the advancing troops would turn and run, the Spartans would then advance at a slow step, and then to a more rapid march, always staying linked together. As the first wave of Persians approached the line of Spartans and their allies the first thing the troops would have seen was a row of nearly identical round shields, each bearing the same sign, the Greek lambda, standing for the Spartan state of Lacedaemon.

In contrast to the heavily armored hoplite, the Persians wore comparatively little armor. In the plains of Asia where mobility was all-important, the Persian was well suited. But they were not prepared to fight shield against shield and sword on sword in the pass at Thermopylae. Only the famous Immortals, the 10,000 men comprising Xerxes' personal elite guard wore anything approaching the armor of the hoplite.

The first wave of 20,000 Medes were repulsed by Leonidas with extraordinary losses on the Persian side. At midday, the Medes were withdrawn and a similar number of Persian soldiers charged the Spartan line but it did not break, and in fact, advanced some yards over masses of dead and wounded Persians. The withdrawal of the Medes had allowed Leonidas to bring fresh troops to the front where they would be interspersed with the 300 Spartiates.

As the afternoon waned and evening approached Xerxes was impatient to get through the pass before dark. It was time for a breakthrough and the Immortals were ordered forward. The Immortals, as mentioned, were the professional elite troops of the Persians and could be compared to the Janissaries under Sulieman the Magnificent or the American Green Berets. But despite their training and dedication, their arrows could not pierce the bronze shields, their short spears could not penetrate the formidable line of the Greeks, and at close fighting they were doomed to be killed or wounded.

It was during the attack by the Immortals that the Spartans initiated a tactic described by Herodotus as follows:

> *One of the feints they (the Spartans) used was to pretend to turn and fly all at once (from the Immortals). Seeing them apparently taking to their heels the barbarians pursued them with a great clatter and shouting;*

> *whereupon, just as the Persians were almost upon them, the Spartans wheeled and faced them. And in this about-turn they inflicted innumerable casualties upon them. In doing this the Spartans had some losses too, but only a few. In the end, since they could make no headway toward winning the pass, whether they attacked in companies or whatever they did, the Persians broke off the engagement and withdrew. It is said that Xerxes, who was watching the battle from his throne, three times sprang to this feet in fear for his army.*

On the following day, August 19, Xerxes threw in thousands of fresh troops and during the entire day wave upon wave threw itself against the Spartans and their allies at the Phocian Wall. To judge from an observation of Herodotus, it seems that by the second day Persian morale was so low that many of the Persian soldiers had to be driven forward by the whips of overseers. The only way to reach the bristling Spartan line was by walking over yards of slain and wounded soldiers slippery with blood.

The second day also ended in failure for the Persians. Leonidas had not been moved one foot from his initial position, and it is estimated that between 15,000 and 20,000 Persians lay dead. It was after this second day that Xerxes and his generals determined that they could not succeed in a direct onslaught against the Spartan line and they sought a way around the mountain so as to attack the Greeks from the rear. Herodotus tells the story of how a Greek known as Ephialtes led the entire contingent of Immortals over Mt. Kallidromos on the evening of the second day.

Now Leonidas had expected a Persian effort to outflank his forces at Thermopylae and he had stationed a guard of about 1,000 Phocians on Mt. Kallidromos to resist any effort of the Persians to come over the mountain. But the Phocians were no match for the Immortals, and in fact Herodotus tells us that they were even surprised by them and ran from battle. Apparently, several of the Phocian soldiers ran to notify Leonidas that the Immortals were approaching from behind. A war counsel was quickly called in the early morning of the third day and Herodotus describes it as follows:

> *Some urged that they must not abandon the post, others the opposite. The result was that the army split, with some contingents returning to their various states while others prepared to stand by Leonidas. It is said that Leonidas himself dismissed many of these in order to spare their lives. As for the Spartans, it would not be in their code for them to desert the post which they had been entrusted to guard.*
>
> *The force that stayed would have numbered less than 2,000 men.*

After the war counsel those defenders remaining paused to eat before the attack came. Herodotus tells us that the salutation of Leonidas to all of them was: "Have a good breakfast men for tonight we dine in Hades."

A number of stories grew up around the Battle of Thermopylae and were told even at the time. They have been recounted by both Herodotus and Plutarch. One story tells of a Spartan who suffered from inflammation of the eyes and had been sent by Leonidas to the camp outside the east gate to recuperate. He returned home with the other Peloponnesians who did not fight on the last day. On his arrival in Sparta, he was received with silent scorn and was effectively ostracized. His only redemption came by his death at the Battle of Plataea. The other story is about a Spartiate who had been sent by Leonidas to seek reinforcements and thus missed the battle. On his return to Sparta and despite a very legitimate excuse, he was treated with such contempt that he hanged himself. The Spartan code was very unforgiving.

On August 20, the third day, Xerxes renewed the attack at dawn, and it raged almost unceasingly for two hours. The Greeks, apparently knowing that they would not survive the day, fought with such fury and ferocity that they actually drove the Persians back. By then the Greeks had lost most of their spears and were fighting with swords hand to hand. It was at this time that Leonidas was killed, and a savage battle developed over the king's body, the Persians being determined to seize it as a trophy. Four times the Greeks drove the Persians off and managed then to carry the King's body back within their ranks.

Then, at mid-morning, came the cry they had expected, a shout that the Immortals were coming up from behind. Herodotus tells us of the last stand of the Spartans and their allies:

> *They drew back into the narrow neck of the Pass and formed themselves into a compact body altogether and took up their stance on the mound. This is the hillock near the entrance (near the Phocian Wall) where now stands the stone lion in memory of the lion's son. In this place they defended themselves to the last, with their swords, if they still had them, and if not even with their hands and teeth. Then the Persians from in front, piling over the ruined wall, and those who closed in from behind, overwhelmed them.*

The Persians searched and found the body of Leonidas. He was beheaded and his head was displayed before the Persian troops on a pole.

The story of the rest of the war cannot be told here. The Persian army and navy moved south. Athens was occupied, and in September the Battle of Salamis was fought in the narrow strait between the Greek mainland near Athens and the Island of Salamis. That battle was a complete victory for the Greeks. Xerxes then took the main body of his army and his navy and moved north and across the Hellespont before the end of the year, leaving Mardonius and an occupation force of some thirty to fifty thousand men to keep the Greeks subdued. The following summer, near a small town no more than ten miles from the city of Thebes, the Persian occupation force met a Greek force of equal size, this time led by nearly 8,000 Spartiates. The result of the Battle of Plataea was a rout of the Persian forces and an effective end to the war.

Following the war, the League of Northern Greek States, whose meeting place had always been at Thermopylae, set up a simple plaque to commemorate the last stand of Leonidas and his men. Its brevity and directness were typically Spartan.

Go friend, and tell Lacedaemon,

That here, obedient to her laws, we lie.

From a strategic point of view, the defense of the Pass at Thermopylae permitted the Greek navy, using Artemisium as its base, to inflict significant losses on the Persian fleet. You will recall that Themistocles had moved the Greek fleet to the Artemisium on the north end of Euboea, in coordination with the defenders at Thermopylae. In fact, it was on the final day of the Battle of Thermopylae, August 20, that Themistocles chose to confront the Persian fleet. Leaving Artemisium with some 300 ships the Greek navy challenged nearly 450 warships of the Persians to what naval historians agree was a draw. Although a stalemate, psychologically the sea battle at Artemisium proved an invaluable experience. According to Plutarch, it shattered the myth of Persian naval superiority, and the Greeks realized they could hold their own against an unfamiliar and larger enemy. Immediately after the Greek fleet returned to Artemisium, it learned of the Persian breakthrough at Thermopylae, and Themistocles quickly moved the fleet south to Salamis.

The death of Leonidas, and of the Spartans and their allies at Thermopylae also served to unify, if only for a very short time, a divided and irresolute Greek people. The 300 Spartans had died to a man and a king had been sacrificed not in some local war in the Peloponnese but far to the north in a battle to preserve freedom for all of Greece. Thermopylae became, almost immediately, a symbol to all Greeks of pride and honor.

The Greek triumph over the Persians also ushered in the so-called golden age of Athens - a 50 year period in which architecture, art, literature and democracy flourished, a time when Aeschylus, Sophocles, Pericles, Phidias and Socrates were at their zenith. No historian doubts that without that victory the rise of Athens could not have occurred. Rather, it would have been a vassal state of the Persian Empire and its fledgling democracy would have withered and died.

Finally, to return to the ideas expressed in the earlier quotation from Montaigne, the Battle of Thermopylae has come to symbolize what is most noble in the human spirit. If the Greek victory over the Persians was seen as the triumph of freedom over tyranny, of individualism over collectivism and of democracy over autocracy, then Thermopylae has also become symbolic of this triumph. The single minded and selfless heroism demonstrated by the small force of Greeks at the pass of Thermopylae has been celebrated through the ages. The fact that the Greeks lapsed into war among themselves within 50 years still does not tarnish the fact that a few of them at Thermopylae demonstrated the highest degree of courage, idealism and self-sacrifice. It is unfortunate but understandable in human terms that these virtues are not sustainable but for a short period of time.

William Golding has captured some of the essence of what the Battle of Thermopylae has meant to all of us in an essay entitled "The Hot Gates":

It is not just that the human spirit acts directly and beyond all arguments to a story of sacrifice and courage, as a wine glass must vibrate to the sound of the violin. It is also because, way back and at the hundredth remove, that company of Spartans *stood in the right line of* history. A little of Leonidas *lies in the fact that* I *can go where* I *like and write what* I *like.* He *contributed to set* us *free.*

BIBLIOGRAPHY

A. R. Burn, *The Penguin History of Greece*

J. B. Bury, *History of Greece*

Ernie Bradford, *The Battle for the West: Thermopylae*

Peter Green, *Xerxes at Salamis*

Herodotus, *History of the Greek and Persian War* (W.G. Forrest, ed.)

The Round Table Presentation | May 5, 1994

The Essential Hero
Admiral Horatio Nelson

It was dawn of October 21, 1805. The Admiral's Log read "Light westerly wind, long swells, good visibility." He had spent the past twenty-four hours maneuvering his fleet of 27 ships-of-the-line into a position about nine miles west of the combined French and Spanish fleet, guessing, quite correctly, that this position would give his fleet the favorable westerlies. The combined French and Spanish fleet of 33 ships-of-the-line had left its home port of Cadiz the day before, intending to sail into the Mediterranean. Now it found itself that fine October morning midway between Cadiz and the Straits of Gibraltar, off a low promontory of land about sixty-six feet high called the Promontory of Caves - in Arabic, Tarafelagar. At 6:00 a.m. the British could see the sails of the enemy fleet outlined in the rising sun while they remained undetected in the darkness on the western horizon. Admiral Horatio Nelson signaled his order to form two lines of battle and to advance with all speed. The battle of Trafalgar, the Royal Navy's and Nelson's finest hour, had begun.

My purpose in this paper is to provide a brief history of Horatio Nelson's life, an ambitious project, I admit, given the fact that he served in the Navy for 35 of his 47 years, fought innumerable battles at sea and in marine assaults and sieges, not to mention four major fleet actions, each of which was significant enough to be ranked among the great sea battles of all time, lost an arm and an eye, carried on an open and scandalous love affair with the wife of a close friend, and became, during his lifetime and doubly so after his death, one of England's most cherished heroes. Hopefully in the process, I can provide you with a glimpse of the British Navy during the Napoleonic wars, the ships, the weapons, the strategies and tactics employed.

Horatio Nelson was born in 1758, the sixth of eleven children of the Reverend Edmond Nelson, the Rector of Burnham Thorpe, a village located on that part of the Norfolk coast facing the North Sea. Young Horatio was a small-boned and undersized youth, on the face of it, according to his uncle, not a likely candidate for a life at sea. But Nelson himself had petitioned his relative to take him to sea, and at age twelve young Horatio was listed on the books of the sixty-four gun warship, the Raisonable, as a midshipman. The ship was captained by Horatio's uncle, Maurice Suckling, who later became Comptroller of the Royal Navy and thus could ensure that Horatio received good appointments. Nelson's early experiences at sea were varied, including an expedition to the Artic searching for a Northern passage to the Pacific and service on a frigate around the Cape of Good Hope to India.

In 1777, Nelson's 19th year, he passed his examination for lieutenant, following which he sailed for the West Indies. He was now fully grown, 5 feet 5 1/2 inches tall, slightly built and generally considered handsome, with large eyes, arching eyebrows and sandy hair. He was seen as alert and lively, engaging, and ambitious for action and glory. In June 1779 while in the West Indies, he was appointed to the position of Post Captain, which meant that he was rated capable of commanding a ship that carried more than twenty guns. The previous captain had been killed. Nelson wrote at the time, "I got my rank by a shot killing a post captain and I most sincerely hope I shall, when I go, go out of the world the same way." Prophetic words. He was not yet 21, but even at such a young age Nelson was "made"; that is, under the rules of strict seniority then governing the Royal Navy, no junior officer could be passed ahead of him. If he could simply stay alive and not foul up, he would attain flag rank.

The years between 1780 and 1787 were spent primarily in the West Indies where on the island of Nevis he met, courted and within two years married, a young widow named Frances Nisbet. Frances, or Fanny, had a small child, Josiah, whom Nelson adopted and who later would accompany him aboard many of the ships he commanded. Later, when their marriage foundered, much would be made of Fanny's dignified loyalty to the end, but even Nelson's early letters to his new bride are formal and somewhat cold, suggesting that they may have had little in common from the outset.

In 1787 Nelson was recalled to England. Europe was, for a brief time, at peace, and England had no need for a large or active Navy. Nelson and Fanny returned to live with his father in Burnham Thorpe where Nelson attempted fairly unsuccessfully to settle into the life of a country gentleman and had to be sustained by a pension provided him by relatives.

By 1793, the early enthusiasm felt in England for the French Revolution had disappeared, in large part because of the actions of the National Convention in Paris to abolish the monarchy followed by the execution of Louis XVI. As a result, England broke off relations with France, and on February 1, 1793 the National Convention declared war on England. Within days Nelson was informed that he would be given command of his first ship-of-the-line, the Agamemnon. His principal responsibility over the next four years was to assist in the blockade of Toulon and Marseilles, Toulon being the principal dockyard for the French Mediterranean fleet. He would also figure prominently in the campaign to take and hold Corsica for England.

In the Corsican campaign, which began in 1794, Nelson distinguished himself and surprised both officers and crew by taking personal command of the marines in the sieges of Bastia and Calvi - two fortified seaports on the island.

It was during the Calvi campaign of 1794 that a cannonball struck the wood and earth fortification directly in front of him and a splinter of wood entered Nelson's right eye, resulting in immediate and nearly total blindness. Nelson was thereafter able to distinguish light from dark, but in all other respects his right eye was useless. From that time on he took to wearing a green eye shade immediately under his hat in order to protect his one good eye from the glare of sun and sea.

During these four years Nelson was engaged in several minor sea battles, and disagreed with the traditional approach taken by his senior officers who seemed to be content to engage the enemy briefly, capture a prize or two, and then withdraw satisfied. Nelson's letters to Fanny are brimming with frustration that full battles did not ensue with a decisive victory or defeat as the final outcome. It was Nelson's quest for total victory that served as the hallmark of the battles in which he commanded.

Meanwhile, this same four year period in France saw the rise of Bonaparte and the extraordinary success of the French army in an Italian campaign that brought France control of all the port cities in that part of the Mediterranean, and forced the British to abandon the island of Corsica. Finally, in December 1796, the decision was made by the admiralty in London to withdraw the fleet from the Mediterranean. Only Gibraltar would remain as the sole English presence in the Mediterranean. Nelson was responsible for the evacuation of Elba and left the Mediterranean barely in time to join Admiral John Jervis in what was to be the battle of Cape St. Vincent, fought on Valentine's Day, 1797. This battle was an interesting anticipation of Trafalgar, being fought in the same waters (in the Atlantic off the southwest corner of Spain) and against a Spanish fleet that was now allied with France.

On a tactical level, this battle has always been famous for an inspired act of disobedience on Nelson's part, without which the British fleet would probably have achieved only a limited success, if any at all. [See Diagram]

Admiral Juan De Cordova sailed into this fleet action with twenty-seven ships of the line. These included the biggest, most heavily armed warships in the world. Cordova's flagship, the Santissima Trinidad, the largest in the world, bristled with 136 guns. In contrast, the British fleet, with only 15 ships of the line, boasted only two 100 gun ships. To naval historians, it was the discipline, training, seamanship and gunnery of the British that accounted for their success. That, and Nelson's intuitive action to engage the enemy in violation of Admiral Jervis' orders.

Daylight on February 14 revealed the Spanish fleet in some disarray, with a large gap separating the Spanish main body from the warships in the vanguard. The Spanish ships were moving to the northeast attempting to make port at Cadiz. The British fleet came upon them in a single straight line of battle from the Northeast making for the gap in the Spanish line. Sailing through the gap and thus dividing the fleet, Admiral Jervis then ordered the entire line of fifteen ships to tack from its southwesterly direction toward the north in order to confront the main body of the Spanish fleet. As the British fleet was executing this maneuver de Cordova ordered the main body of his fleet to turn and cross behind the last ship in the British line in order to avoid a direct confrontation and make their escape.

From his vantage point in the Captain, third from the end of the British line, Nelson saw the Spaniard's intention and knew that if he maintained his position in the line, the Spanish fleet would escape. He therefore abandoned one of Naval warfare's most sacred principles - that the line of battle, once formed, could not be altered. Nelson brought his ship around out of the line and directly confronted the leading Spanish ships, his intention being to slow them down until the rest of the British line could be brought to bear. As it happened, Nelson's close friend, Collingwood in the rear-most British ship, the Excellent, turned and followed Nelson.

Nelson's action took the Spaniards by surprise, and battle was joined. The confrontation of the van of the Spanish fleet by Nelson's Captain and Collingwood's Excellence caused the San Nicholas and San Josef to collide with Nelson's ship, thus enabling Nelson to perform one of the most famous exploits in British naval history, the taking of two ships at one time. Nelson described it this way:

> "*At this time the Captain, having lost her foretopmast, not a sail, shroud or rope left, her wheels shot away, and incapable of further service in the line, or in chase, I ordered Captain Miller to put the helm a-starboard (steering by the Ship's tiller) and calling for borders, ordered them to board.*"

Led personally by Nelson, the seamen and marines of the Captain swarmed onto the San Nicolas and, because the San Josef was locked alongside the San Nicolas, Nelson ordered and led a second boarding attack onto the second ship. Both ships surrendered to him. By the end of the battle the British had captured four ships and had heavily damaged a good many others. Nightfall allowed the bulk of the Spanish fleet to limp away to the protection of Cadiz.

When Nelson went on board the Victory, Jervis' flagship, at dark on February 14, he was warmly embraced by the admiral. And when Jervis' first captain, Robert Calder, pointed out that Nelson's decision was a serious breach of the Fighting Instructions normally calling for a court martial, Jervis, the Navy's strictest disciplinarian, said: "It certainly was so and if ever you commit such a breach, I will forgive you also."

While Jervis was credited with the victory, the role of Nelson was not lost on his countrymen. He was knighted and given the Freedom of the City of London. [But he was not promoted. Promotion came in April in the ordinary course of seniority when he became Rear-Admiral of the Blue.]

But failure and near personal disaster followed hard on the heels of this success. In the summer of 1797 Nelson received orders to proceed to the island of Tenerife off the west coast of Africa and capture the port of Santa Cruz. Attempting to take the town by surprise during the night, the winds and tides refused to cooperate and a coordinated attack failed. In the battle Nelson was struck by a bullet that shattered his right arm, above the elbow. Once aboard his warship the arm was immediately amputated by the ship's surgeon. There was no general anesthetic in those days and Nelson was apparently given a drink of rum and a leather pad to bite on during the operation. Nelson's recuperation took nearly a year during which time he concluded that his naval career was over.

Before moving into the next phase of Nelson's career, let me pause to review the conditions of the British navy during the Napoleonic wars, the ships, arms and men, and to shed some light on how Britain was consistently able to assert its supremacy over the French fleet.

Let me begin with a description of the ships. The primary vessel of war was the full-rigged sailing ship - the man-of-war. Its design had essentially remained the same for over 200 years. The British man-of-war was built of English oak, the largest ships requiring more than 3,000 loads of timber, the equivalent of what would be grown on forty acres of land in 100 years. The ships were triple-masted, the masts normally being of single trees of Scandinavian pine.

The warships were classified according to the armament they carried. The largest ships in the British Navy, having between 100 and 120 guns, were labeled "first rates". Those ships of 90-98 guns - second rates, ships of 64-84 guns - third rates and so on down to "sixth rates" which carried 20-28 guns. The first, second and third rates were denominated "ships-of-the-line" because, being the most heavily armed and the most powerfully built, they were the ships ordinarily used to form the line of battle. The fifth and sixth rate ships were frigates, lighter and sleeker than the ships-of-the-line and thus could outsail them. Frigates were primarily used as scouting and messenger ships.

Nelson's flag ship at Trafalgar, the Victory, was a "first rate". It was 186 feet long, 52 feet at its broadest point, and displaced approximately 3,500 tons. Its main mast stretched 205 feet above the waterline, and it employed over 90 tons of ropes used to hoist and furl the thousands of square feet of canvas sails. The Victory, as with other men-of-war, was sheathed with copper below the waterline to protect it from the teredo worm, a voracious mollusk that could eat its way through the ship's timbers, and leave a honeycombed hull that could easily give way and cause the ship to sink.

Ships-of-the-line might make a top speed of approximately seven knots, whereas a frigate could make as much as nine knots per hour. Speed was not as important as position vis a vis the wind, for the tactical objective in a sea battle was to have the wind astern to enable tacking to bring the broadsides to bear. This was referred to as the "weather gauge."

During the Napoleonic wars and particularly at the time of Trafalgar in 1805 England had 175 ships of the line and 246 other warships for a total of 421 war vessels. Approximately two thirds of this number were in commission at any given time. France had fewer warships; however, after France allied with Holland, Spain and Denmark, their combined fleet was greater than that of the English fleet. French ships of the line were on the whole better designed than the British. They had a broader beam which made them steadier in heavy seas (you have to picture firing a cannon that is rocking through thirty degrees of arc). The French design also had the gun ports higher above the waterline so that the lowest level of guns could be used in rougher seas.

The primary weapon used on a man-of-war was the cannon. It was normally made of iron and was muzzle-loaded. Although breech loading was understood at the time, breeches had not been developed that were strong enough to withstand constant firing. The cannon were smooth bore using gunpowder as a propellent. The maximum range of the balls was approximately 2,500 yards, although 400 yards was the preferred firing range. At that distance a ball could penetrate three feet of timber and the splinters generated from impact were often lethal.

The primary damage inflicted on enemy warships was from the pounding made by these solid cannonballs. Rarely was a ship ever sunk. The primary effect of cannon fire was to disable the ship by destroying masts and rigging. The ship could then be taken by boarding. Although powder-filled balls had been experimented with by the French, it was deemed too dangerous to use on a warship. The heaviest guns fired 42 pound balls, although 32 pounders were preferred because of the higher efficiency with which they could be loaded and fired. In addition, the cannons fired chain-shot consisting of two half balls attached by a chain, and this was used primarily to destroy the masts and rigging of enemy ships. At close quarters grapeshot or canister was used. Each cannon was manned by a crew of fifteen which included the powder monkeys, usually boys who were required to go to the powder magazine to bring up gunpowder, load it into flannel bags in a preparation area and make it ready for firing.

Aiming of these cannons was primitive. They had almost no ability to be moved either up or down or side to side. Gimbals or other stabilizers were not yet in use that would allow the cannon to remain relatively level despite the heaving and pitching of the seas so that in rough seas having the cannonball hit anything was oftentimes a matter of pure luck. After each firing, the cannon would recoil several feet. The barrel would then be cleaned, gunpowder rammed home, followed by the ball. Blocks and tackle would then be used to haul the cannon forward to the gun port. It was a ponderous process but a good crew could fire three rounds in less than two minutes. The British rate of fire was approximately three times as fast as the French.

Immediately prior to an engagement, sand and ashes would be scattered around each cannon. Its grim purpose was to assure crews of good footing when the decks became slippery with blood.

The size of the crew on the Victory at the time of the Battle of Trafalgar was 837. Approximately 600 of these men served as gun crews, sailors, cooks and stewards and 85 of the 600 came from a dozen other nations, most of whom had been impressed into service from merchantmen accosted on the high seas. The other 237 consisted of officers and marines, the last for sharpshooting, boarding enemy ships and shore landings.

Approximately one-third of the crew in a typical British man-of-war were volunteers. The remainder had to be coerced to join the Navy primarily by legal process. The result was that many of the ship's crew were sentenced criminals who were ordered to serve in the Navy rather than in jail. You might remember Dr. Johnson's statement that "No man will be a sailor who has contrivance enough to get him in jail; for being in a ship is being in a jail with the chance of being drowned."

Living conditions below decks were, for us, hard to imagine. There was little room to sleep or eat. There were only four heads for 800 men. Descriptions of the conditions below decks in a storm were as disgusting as you might think. Disease was rampant. Dampness was everywhere. During the Napoleonic wars, many English ships on blockade duty were at sea constantly for more than a year. Given this fact it is no surprise that the officers and crews of British ships had so much more training and experience than their French or Spanish counterparts, and this gave the British an important edge in battle.

Now for a brief consideration of strategy and tactics. At the strategic level, it was essential for Britain to keep open its sea routes to sustain its economy. Moreover, England well knew after 1797 that Napoleon wished to invade England and knew that France would have to have control of the sea in order to do so. Because Napoleon and the French allies could harass British shipping anywhere and everywhere, England evolved a strategy of the so-called "close blockade". It was England's intention to keep the fleets of France and its allies bottled up in their ports, primarily Toulon, Cadiz, Ferrol, Brest and Boulogne. If these fleets were to venture out of their ports, the British objective was to immediately attack and destroy them. The French conveniently adopted a complementary strategy. Its mission was to keep its fleets as a constant threat to England, to come out of port on their own terms and to fight the English at a time and place of their own choosing. As a continental power, they were not so dependent upon shipping to support a wartime economy, and their focus was military rather than naval.

On a tactical level, the objective of either fleet could be simply stated but its successful execution was an enormous challenge. The primary objective was to obtain the position of favorable wind (the so-called "weather gauge"). Putting the fleet to windward allowed it to choose the time, direction and method for confronting the enemy. Fights were ship versus ship duels usually sailing on parallel courses, often at point blank range. At the time of the defeat of the Spanish armada, Naval battles were melees, all confusion without coordinated plans of attack. Over the next 200 years a common tactic had developed among all the great navies - the single line of battle. The line of battle was closely drawn so as to prevent enemy ships from dividing it. The hope was to catch the enemy with its ships in some disarray - as at Cape St. Vincent - and thus divide the enemy without in turn having one's own line of battle pierced, then to fall upon one part of the divided enemy with a superior force.

Any yachtsman can appreciate the difficulties in taking even a small boat out the harbor under sail alone. What is so hard to convey today is the challenge in taking one of these behemoths into battle. The manpower just to tack a warship was enormous, requiring skill and strength and involving scores of men in a highly complex series of maneuvers. It is doubly hard to imagine directing the coordinated movements of an entire fleet.

Finally, the maintenance of order and discipline aboard a warship required constant attention. There is no doubt many captains were free with their use of the lash and that cruelty existed on many of these ships. The tyranny of many English captains led to numerous mutinies in the late 1790's. One of the most inhuman tortures was that of 300 lashes, called "flogging round the fleet" which few, if any, ever survived. As an aside, Captain Bligh, of mutiny on the Bounty fame in 1789, served under Nelson with distinction at the Battle of Copenhagen twelve years later in 1801.

The fact was Nelson and his "Band of Brothers" as he called those captains he fought with at the Nile and Copenhagen, were not of this ilk. They were humane in their treatment of the sailors under them and were beloved by their crews. An example of how the crew of one of Nelson's ships felt about his leadership occurred during the so-called time of the mutinies in 1797. Nelson, newly an admiral, was transferred to the Theseus, a ship in which there had been wide-ranging mutinous conduct. He took with him a captain Miller. Within six weeks he found a paper that had been dropped on the quarter-deck that read:

> *"Success attend Admiral Nelson God bless Captain Miller we thank them for the officers they have placed over us. We are happy and comfortable and will shed every drop of blood in our veins to support them, and the name of Theseus shall be immortalized as high as Captain's (the Captain was Nelson's ship at the Battle of St. Vincent).*

Ship's Company

In the spring of 1798, Nelson was again recalled to service with the British fleet off Gibraltar. Word out of France was that Napoleon was on the move looking to the eastern Mediterranean for new lands to conquer, perhaps even Egypt. On May 19, 1798, led by Napoleon, the Army of the East set sail for the invasion of Egypt. Four Hundred troop ships comprised the armada escorted by thirteen French ships-of-the-line, eight frigates and a number of Venetian warships. Nelson, then a rear admiral of the blue, the lowest flag rank, was chosen to lead a fleet of fourteen warships into the Mediterranean to intercept the French. Buffeted by a violent storm that caused Nelson to lose his frigates (and thus his eyes and ears), Nelson and his fleet sailed eastward and discovered that the French army and navy had taken Malta and after a brief stay had departed. In consultation with his captains Nelson guessed that the French might be making for Egypt and he set after them. Sailing into the eastern Mediterranean, the British fleet missed the French fleet by only a few hours on two separate occasions and in fact Nelson arrived off Alexandria on June 29, the day before the French fleet arrived. Without realizing that his guess was correct, Nelson concluded that Napoleon must have had other plans, and he continued his search, sailing along the coast of Palestine, Asia Minor and back to Sicily. It was then that he learned the French had in fact landed in Egypt where Napoleon had disembarked with his troops, defeated the Mameluke Army at the Battle of the Pyramids and occupied Cairo.

Nelson then turned back to Egypt, using the time in passage to confer at length with his captains and develop a battle plan should they confront the French fleet. These captains were by established reputation, some of the finest in the Navy, this "Band of Brothers", including Troubridge in the Culloden, Foley in the Goliath, Miller in the Theseus, Ben Hallowell in the Swiftsure, and James Saumarez in the Orion. On August 1 Nelson sighted the masts of the French fleet east of Alexandria in Aboukir Bay. The French commander, Admiral Brueys, had been at anchor for three weeks along a line simulating a line of battle, but with the ships approximately 150 yards apart. This meant that the ships were too spread out to provide themselves with mutual support and it allowed Nelson's ships to cut in between them and thus to break this stationary line of battle. Brueys' error was compounded by anchoring the fleet at such a distance from the shore so as to allow English ships to range inside the French line and to attack them from the landward as well as the seaward side. However, with shore batteries to help him, Brueys felt confident enough. Since it was late afternoon, Brueys did not expect an attack until the following day. But Nelson had decided otherwise. [See Diagram] He determined to attack immediately and at about 6:30 p.m., with the light fading, the Goliath led the British assault followed by Zealous, Orion, Theseus, and Audacious, all of which cut inside the French line. Nelson's tactics, laid out with his captains beforehand, was not to follow the fleet's normal practice of ranging his fleet in a ship-for-ship gun duel but to concentrate his force on a part of the French line, in this case by literally surrounding the ships in the French van. Nelson, in the Vanguard, engaged the French line from the seaward side.

At the center of the French line was the flagship, L'Orient, a massive ship of 120 guns and a compliment of 1,000 men. The first British ship to set upon L'Orient, the much smaller Bellerophon was forced to retire. Its place was taken by two other British ships who concentrated fire on L'Orient. In this battle Admiral Brueys was wounded, losing both legs to a single cannonball, but he continued to direct the battle, seated on an arm chair with tourniquets on the stumps. At about nine in the evening L'Orient caught fire and the fire spread rapidly. At about 10:00 p.m., the fire reached L'Orient's magazine and it was blown out of the water in a tremendous explosion that could be heard in Alexandria twenty miles away.

Fighting continued into the morning, described by those present as an eerie and nightmarish spectacle, until the final French ship struck its colors at approximately four a.m. Two French ships of the line and two frigates managed to escape. Of those remaining, one had been sunk, nine had surrendered and two were complete wrecks. On the British side, although casualties had been heavy, not a single ship was irreparably damaged and none had struck its colors. The Battle of the Nile, was, no pun intended, a battle of annihilation. As such, it introduced a new approach to Naval warfare, one whose objective was the utter destruction of the enemy fleet, and one which Nelson was to use in the Battle of Copenhagen and at Trafalgar. The immediate result of the battle was that it marooned Bonaparte and his army in Egypt and it reestablished English control of the Mediterranean. Although Bonaparte managed to escape to France in a French frigate, his Army of the East was forced to surrender following the subsequent British invasion of Egypt in 1801. Nelson's victory at the Nile was a tremendous boost to British morale which had suffered from the French triumphs on land over the preceding years. His victory had eliminated the French Mediterranean fleet. England gained access to the Court of Naples where support had been wavering before the battle. Malta was seized from the French. At home, Nelson's inspired action at Cape St. Vincent was seen as no fluke. He was now a thoroughly popular hero, receiving honors from England and abroad and becoming Lord Nelson of the Nile.

But by far the most dramatic personal consequence of his victory at the Nile was his relationship with Emma Hamilton, wife of the British Minister to the Court of Naples, Sir William Hamilton, which lasted for the last seven years of Nelson's life. This affair, which was open and notorious in every sense of the word, was the stuff of gossip, cartoons and numerous stories at the time. With the advent of movies, Hollywood retold the story and it certainly could be the subject of a separate Round Table presentation.

Arriving in Naples from the Nile in September, 1798 and intending to stay only five days before returning to service, it seems that Emma Hamilton had very different plans. Putting Nelson up at Sir William's townhouse in Naples, Emma took it upon herself to nurse Nelson back to health. He had received a head wound at the Battle of the Nile and was apparently in a frail condition at the time. In this early autumn of 1798 Emma was thirty-three years old, Sir William sixty-eight and Nelson would be forty within a few days.

The five day stopover continued for nearly two years, first in Naples and then, to escape the French army as it moved south through Italy, in Palermo, Sicily. Nelson did not return to England until the summer of 1800, and then did so in the company of Emma and Sir William. It appears that sometime early in 1799 the two had become lovers; certainly that would have been the case by the end of that year for in the subsequent year Emma gave birth to a daughter she named Horatia.

Amazingly, this menage a trois did not seem to disturb Sir William who surely knew that Emma had taken Nelson as a lover. In fact, until his death in 1803, Sir William considered

Nelson to be one of his closest friends.

After Nelson's return to England the little that was left of his relationship with Fanny disintegrated rapidly. In the several formal appearances they made together, it was clear to observers that Nelson had developed almost an open contempt for her. The histories I have read uniformly agree that Fanny had remained devoted to her husband. Without any fault of her own she was simply displaced by the beautiful, strong-willed and sexually voracious Emma. This treatment of Fanny exposed the meanest aspect of Nelson's character and is inconsistent with his otherwise high standards of behavior. Without apology, and ignoring the advice of numerous friends, Nelson, Emma and Sir William took up residence together. Yet while polite society treated his affair with great distaste, Nelson's popularity continued unabated among the greater part of the populace. Nelson returned to the sea in January 1801 when he was assigned as second in command to a fleet in the Baltic Sea under the command of Admiral Hyde Parker. Britain was then confronted by a League of Armed Neutrality established by the maritime nations of northern Europe, including Russia and Denmark. These nations who were trading with France and France's allies, had become exasperated with the interference of the British with their shipping. Demand was made by England for Denmark to withdraw from this league, but the terms of that request were rejected. As a consequence in April 1801 the English fleet attacked the Danish fleet then in the harbor at Copenhagen. The result of the battle was a victory for the British but the battle was in doubt for many hours and in one important respect was notable because of Nelson's decision to ignore a direct command from Admiral Parker. At the height of the action, Parker, who was stationed a safe distance off-shore with several ships from where he could watch the battle, felt that the Danish had obtained the advantage and he signaled the order for a cease-fire and withdrawal. Nelson and his officers all saw the cease-fire signal but Nelson chose to ignore it. The story is told that

Nelson turned to his captain and said, "You know, Foley, I have only one eye - I have a right to be blind sometimes;" and then, putting his glass to his blind eye he exclaimed, "I really do not see the signal!" The ensuing victory put Nelson in effective command of the Baltic fleet and it was he and not Parker who negotiated and signed the cease-fire with the Danes.

Following the Battle of Copenhagen Nelson returned to England and lived on an estate which he shared with the Hamiltons and then with Emma after Sir William died. In 1804 Nelson returned to sea as Commander in Chief in the Mediterranean, which brings us to his final battle and final victory, that of Trafalgar. The battle was in fact the climax of an extraordinary six-month campaign the like of which had never been seen before or since. In that campaign, Napoleon, using the Mediterranean and the North Atlantic as a chessboard, sought to assemble a superior naval force in the English Channel to allow the French invasion of England.

The elements of Napoleon's plan involved the French fleet in Toulon, commanded by Admiral Villeneuve and which was blockaded by Nelson, to leave Toulon and join with the Spanish fleet of Admiral Gravina in Cadiz and with Admiral Ganteaume's fleet in Brest both of which were also blockaded by the British. Villeneuve and Gravina were to join up in Cadiz and sail to the West Indies where they would collect all available French warships located there. Ganteaume was to escape from Brest and join forces with the combined fleet in the West Indies. The resulting massed fleet would then recross the Atlantic and sweep the English Channel for Napoleon's invasion. Villeneuve and Gravina did meet and sail to the West Indies with Nelson and his fleet in pursuit. The combined fleet waited to be joined by Ganteaume but after more than a month of waiting, Villeneuve acknowledged that Ganteaume must not have been able to get past the blockade to join his fleet. Accordingly, Villeneuve headed back to the Coast of Europe with Nelson only several days behind. In August, Villeneuve took his combined fleet into the port at Cadiz and Nelson laid off the coast nearby in wait for Villeneuve and his fleet to come out.

In the meantime, Nelson had contrived a strategy for what he perceived as the inevitable battle. Instead of forming his fleet into a single line of battle, he chose to put it in two parallel lines to pierce the line of battle of the combined fleet and fall upon the ships of its center and rear. At several dinner parties aboard his flagship (that served as briefings) he outlined his plan to the other captains, and all were enthusiastic. On October 19, Villeneuve left Cadiz with his fleet and the 19th and the 20th were spent in maneuvering, with Nelson pulling off into a position some nine miles to the west of the combined French fleet. By first light on the morning of October 21, Villeneuve realized that his chances of getting clear of the British fleet unmolested had disappeared, and he signalled his ships to reverse course and make for Cadiz to the northwest. But it was too late. Nelson and his close friend and second in command, Admiral Collingwood, had already placed each of their flagships (Nelson in the Victory and Collingwood in the Royal Sovereign) at the heads of two British columns driving like two slow-motion lances at the center of the combined fleet, twenty-seven British ships-of-the-line against thirty-three of the combined fleet.

At 11:45 a.m. as the two columns approached within a mile of the combined fleet, Nelson hoisted his famous signal, "England expects that every man will do his duty". Collingwood's reaction in Royal Sovereign to the south of Nelson was to mutter "I wish Nelson would stop signalling. We know well enough what we have to do."

There was one very unpleasant aspect to Nelson's plan. Because the British ships were coming straight at the line of the combined fleet, the lead ships in each column, in this case Nelson's and Collingwood's, took the brunt of broadside fire from the enemy during the last mile before the ships engaged. For this reason both admirals had assented to the wishes of their captains the day before that they not place their ships at the head of each column, but further back where the murderous initial barrage would not be felt. Yet when the attack began, both admirals ignored their agreement. In fact Nelson and Collingwood literally raced each other to be the first ship to engage the enemy. Collingwood was the first to come under fire.

Nelson kept his frigate captains aboard Victory until he was certain that his tactic would succeed and his columns would pierce the enemy's line of battle at two points. He then released the captains to speed down the columns and signal all ships that "if by the mode of attack prescribed they found it impractical to get into action immediately they might adopt whatever they thought best provided it led them quickly and closely alongside an enemy." This was representative of Nelson's style of leadership - flexibility, willingness to delegate broad authority to his captains, and implicit trust in their skill and judgment.

Collingwood's Royal Sovereign and Nelson's Victory came into direct confrontation at about noon with the other ships in each line following and seeking out adversaries for ship to ship battles. At 1:15 p.m. in a murderous fight with the Redoubtable, a sharpshooter from the masts of the French warship picked off Nelson as he paced the Victory's quarter-deck. The bullet went through Nelson's chest and lodged in his spine. The ball had severed a main branch of the pulmonary artery. He was carried below decks and made comfortable. All knew he was dying. At about 4:15 Nelson was informed by his captain, Thomas Hardy, that the battle had gone well and after assuring him that no British ship had struck its colors, Nelson, lucid till the end, spoke his last words: "Now I am satisfied, thank God, I have done my duty." He died at 4:30 PM. At that time the British victory was complete.

By any measure, Trafalgar was a sea battle of superlatives. Eighteen of the combined fleet were either taken as prizes or destroyed (sinking in the gale that followed), while the British lost only a single ship. It proved the ruin of Napoleon's naval ambitions as well as the end of any real hope of invasion of England. Of those ships that escaped the destruction at Trafalgar, eleven that reached Cadiz never put to sea again.

At least three things were responsible for the British victory: the superiority of British seamanship, sharpened by the months of convoy duty, the superiority of British gunnery, and Nelson's genius in devising an unorthodox manner of attack and his ability to obtain the enthusiastic support of his captains to carry it out. While Nelson was possessed of intelligence, insight, charm, personal magnetism, extraordinary powers of expression and persuasion, fierce determination, devotion to duty and undeniable courage, no less important to his success were his skill and insight in choosing subordinates, his willingness to take them into his complete confidence, to communicate effectively to them, to encourage and praise them and give them the fullest credit possible, but in failure to take all blame upon himself, a trait that Robert E. Lee possessed in like measure. History seldom finds such a generous spirit in a position of high command.

Trafalgar was the greatest of Nelson's three classic victories (the others being the Nile and Copenhagen) and it was the last fleet action of the age of sail to be fought on the open sea.

The extent of Nelson's immortality is evident in this eloquent tribute paid by Captain Alfred T. Mahan of the United States Navy:

> *The words, "I have done my duty," sealed the closed book of Nelson's story with a truth broader and deeper than he himself could suspect . . . . Other men have died in the hour of victory, but for no other has victory*

> *so singular and so signal graced the fulfillment and ending of a great life's work . . . . There were, indeed, consequences momentous and stupendous yet to flow from the decisive supremacy of Great Britain's sea power, the establishment of which, beyond all question or competition, was Nelson's great achievement; but his part was done when Trafalgar was fought. The coincidence of his death with a moment of completed success has impressed upon that superb battle a stamp of finality, and immortality of fame . . . . He needed and he left, no successor. To use again St. Vincent's words, "There is but one Nelson."*

In the words of historian Oliver Warner:

"Trafalgar marked the end of the career of the most illustrious admiral in his country's history, fulfilled his wish for the annihilation of his opponents as a coherent force, and secured for the Royal Navy a supremacy which was unchallenged for more than a century.

That was enough for one autumn day."

BIBLIOGRAPHY

Great Battle Fleets, Oliver Warner; 1973 The Hamlyn Publishing Group Limited.

Naval Warfare, An Illustrated History, Richard Humble, Ed.; 1983 Orbis Publishing Limited.

Nelson the Commander, Geoffrey Bennett; 1972 Charles Scribner's Sons.

Nelson, The Essential Hero, Ernle Bradford; 1977 Harcourt, Brace, Jovanovich.

Twenty-five Centuries of Sea Warfare, Jacques Mordal; 1965 Bramhall House, New York.

The Round Table Presentation | April 4, 1996

The Professor From Bowdoin
Joshua Chamberlain

I know from various conversations that many of you have read The Killer Angels by Michael Shaara, an exceptional historical novel about the battle of Gettysburg. One of the central characters in that novel and, of course, in the battle, was a 34-year old professor from Bowdoin College in Brunswick, Maine named Joshua Lawrence Chamberlain. The book only hinted at his background and made a few comments about his life after Gettysburg, and I was fascinated to learn more. What I found was a compelling story about a remarkable man, a story with Gettysburg as its centerpiece, but which by no means ends there. It's a story that involves undisputable acts of heroism. It permits an appreciation of the luck in being in the right place at exactly the right time, and for being in the wrong place at the worst possible moment. And finally, in a larger sense, it is about how one man's acts affected the course of the Civil War and how that war left its indelible stamp on that man.

Joshua Lawrence Chamberlain was born in a one-story cottage near Bangor, Maine on September 28, 1828, the first of five children of Joshua and Sarah Chamberlain. Located some 30 miles up the Penobscot River from Penobscot Bay, Bangor was a booming lumber town on the edge of a great pine forest wilderness. By the time Chamberlain was twelve in 1840, Bangor had become the world's leading lumber port, and for a time, the Penobscot was very likely the busiest waterway in the country.

Chamberlain grew up on a farm and, by his teenage years, had already learned the craft of taxidermy, had learned to play numerous musical instruments, and he had developed a love of reading, although the reading of "novels" was discouraged by his parents. In fact Lawrence, as he was called, was not permitted to read Cooper's The Deerslayer as it was believed by his father to be "too frivolous." Chamberlain was a slender, well-built and graceful, athletic young man, considered by the ladies to be good-looking. Lawrence's father had always hoped that he would go to West Point. But Chamberlain defied his father's wishes and resolved to go to Bowdoin College, just down the coast at Brunswick, Maine. At the time he entered Bowdoin in 1848 he was characterized by his teachers as somewhat shy. He apparently possessed a stammer in times of stress, and was perceived overall to be a studious, intense and ambitious young man.

Bowdoin College in 1848 consisted of four buildings and an unfinished chapel. It had been founded in 1802 with one instructor and eight students, and it maintained a rigidly classical and orthodox curriculum. Even so, at the time of Chamberlain's matriculation, it claimed an impressive faculty and equally impressive graduates, including Franklin Pierce, Henry Wadsworth Longfellow, and Nathaniel Hawthorne.

Chamberlain was a serious and sober student, steering clear of the large group of his classmates who had cultivated a hard-drinking and rough-housing reputation. Despite his aloofness, he was liked by the other students, and when one of the new fraternities sought to have him join, Chamberlain was tempted. At the time he was attending a prayer circle and he sought advice from a faculty member who was also a member of the circle. The faculty member stated that while he knew of nothing in fraternity life that might offend

Chamberlain's religious sensibilities, he did acknowledge that on occasion, "There may be an excess of hilarity and jocoseness at which an exceedingly scrupulous conscience might be a little troubled." Could it be that people really spoke this way? In all events Chamberlain did not join the fraternity.

Chamberlain's career at Bowdoin was marked by a broad-based curriculum including subjects ranging from Old Norse to Advanced Mathematics. He was elected to Phi Beta Kappa, and he joined a circle of young men and women who met every two weeks to present and discuss papers prepared by each. This circle was most appropriately called The Round Table.

While at Bowdoin Lawrence studied Hebrew literature in the class of the new professor of natural and revealed religion, Calvin Stowe, who moved to Brunswick in 1850 with his wife and several children. Professor Stowe's wife, Harriet Beecher Stowe, held Saturday evening gatherings for a group of friends and students of her husband, including Chamberlain. The highlight of these evenings was the hostess reading each installment of a novel she had been writing, Uncle Tom's Cabin.

Chamberlain graduated from Bowdoin with honors in 1852. Although his father encouraged him to attend West Point following his graduation from Bowdoin, Chamberlain decided to seek a graduate degree in religion by way of a three-year program at Bangor Theological Seminary. In 1855 Chamberlain took his degree and received an offer from Bowdoin of a junior instructorship in religion as well as a number of requests from churches in Maine to serve them as pastor. He turned the churches down, deciding that he was more of a scholar than a preacher, and he moved back to Brunswick, taking with him his new wife, Frances (Fanny) Adams, the daughter of the minister at the church he attended in Brunswick while a student at Bowdoin.

In 1856, after just a year at Bowdoin, Chamberlain was promoted to a professorship in rhetoric, a permanent position with an annual salary of $800, and in October 1856 Fanny gave birth to their first child, a daughter. Lawrence and Fanny would have five children in all, but only two would survive to become adults.

In the meantime, Chamberlain continued to teach at Bowdoin and moonlighted by teaching classes in Spanish, Old Norse, and Early English Literature. Bowdoin retained its rigidly classical curriculum and also persisted in an independent, even states' rights, political point of view. As the United States drifted toward disunion, Bowdoin seemed out of touch with the significant events of the day and the prevailing mood in the north. Indeed, in August 1858, a few weeks after Abraham Lincoln's "House Divided" speech to the Illinois legislature, Bowdoin granted an honorary degree to Mississippi Senator Jefferson Davis.

When war came in the spring of 1861, Chamberlain confined his patriotic activities to speaking at recruiting rallies and writing recommendations for Bowdoin graduates who were seeking officer commissions. But he began to feel restless about his own role as a professor and became increasingly interested in seeking a commission for himself. The Bowdoin faculty and administration balked at allowing him to leave, intending to hold him to his contract. In August 1861, Bowdoin appointed him professor of Modern Languages and, at his request, granted him a sabbatical to study in Europe during part of 1862.

> *In July 1862 Chamberlain wrote the governor of Maine requesting a commission in one of the five new regiments being raised in the state in response to an appeal from President Lincoln. This appeal came following the defeat of McClellan's army in the Virginia Peninsula in a series of battles known as "the seven days." Chamberlain's letter included the following:*

> *I have always been interested in military matters, and what I do not know in that line, I know how to learn. I fear this war, so costly of blood and treasure, will not cease until the men of the north are willing to leave good positions and sacrifice the dearest personal interests.*

On August 8, 1862, over a formal protest from the faculty and administration at Bowdoin, the governor commissioned Chamberlain as a Lieutenant Colonel of the 20th Maine Volunteer Infantry.

Within a month Chamberlain embarked with the 20th Maine on a steamer that carried him to the Potomac. In the regiment was Lawrence's 21-year old brother, Tom, who had been serving as a storekeeper and became the Quartermaster Sergeant of the regiment.

No sooner had they disembarked the steamer at Alexandria when they learned that Lee had marched his army into Maryland. The 20th Maine promptly marched northwest to join the Army of the Potomac and arrived in time to serve as a reserve unit for the Battle of Antietam. By 1862, experience had taught that it took at least three to four months to prepare a new regiment for actual combat, so the 20th Maine was not involved in the fighting at Antietam.

In the weeks that followed Antietam, Chamberlain spent as much time as he could reading and studying books on infantry tactics, as well as memorizing much of the Army manual concerning dress, protocol, drilling and so on. The regiment had been outfitted by the state of Maine with heavy woolen garments, probably suitable for Maine winters and tolerable in Maine summers, but wholly inappropriate for the hot climate of Maryland and northern Virginia. The heat and the unvarying diet of hard bread, salt pork and coffee took its toll, and by the end of October 1862, the 20th Maine's sick list contained more than 300 names out of approximately

1,000 in the regiment.

Chamberlain supplemented his book learning by participating in several raids and sorties across the Potomac. In an action near South Mountain in mid-October 1862 against some of Jeb Stuart's cavalry, he was fired upon and a ball tore the cap off his head. Back in camp he wrote Fanny, "Most likely I shall be hit somewhere at some time, but all 'my times are in His hand', and I cannot die without His appointing." In addition to this sense of serene fatalism, Chamberlain's letters also convey with singular force how much he loved the military, relished being part of something that was bigger than life, and enjoyed the authority conferred on him as a Lieutenant Colonel. Unlike Oliver Wendell Holmes who was wounded at Antietam and who characterized the war as "an organized bore," Chamberlain's letters expressed great enthusiasm about himself and his regiment. He realized he had physical courage, and he learned he had a gift of command.

December 1862. The Army of the Potomac was now under the command of Ambrose Burnside, and was camped across the Rappahannock from Fredericksburg. The cold was so intense that on the night of December 6 two members of the 20th Maine froze to death in their tents.

It was at Fredericksburg that Chamberlain saw his first real action in the midst of a great battle. On December 11, Union forces laid down pontoon bridges, crossed the Rappahannock and occupied the town. From there they advanced toward the Confederate position strung out along Marye's Heights behind the town. The Heights were defended by Longstreet's corps. Along the wooded crest of Marye's Heights Longstreet's cannons were placed. Near the base of the hill several hundred yards outside of town ran the road to Richmond, sunken along that stretch and lined on the city side with a shoulder-high stone wall. A near perfect defensive position, Longstreet stationed 2500 riflemen from two of his brigades arranged in ranks four deep behind that wall.

On the morning of December 13, Union forces stormed Marye's Heights. The result was a carnage. The rifle fire from the stone wall was so murderous that out of 14 charges made against it that day by thousands of infantry, not one Union soldier ever reached it. Chamberlain, whose regiment participated in the 13th assault, witnessed the first charge from the town. He recalled that when the volley of rifle fire came from the stone wall, "in an instant the whole line sank as if swallowed up in earth."

He recalled the fearful bloodshed of his regiment's advance. Stopped by the withering rifle fire within about 100 yards of the stone wall, Chamberlain and his men took cover as best they could behind piles of dead bodies. Then as the shooting died down, the survivors confronted a new horror. Sweat-soaked from the exertion of the charge, men began to suffer from the extreme cold. As the night wore on, Chamberlain and other survivors pulled the bodies of the dead over them to keep warm. The next day, not having been relieved, the 20th Maine held off a determined Confederate charge by firing from behind stacks of corpses. When relief came on the following day, Chamberlain and his regiment had been 36 hours on the front line.

The Union front was so undefined on the second night before the stone wall that, near midnight, when Chamberlain picked his way through the bodies to encourage his men, he accidentally came upon an enemy picket. Thinking he was speaking to one of his own men, he said, "Throw the dirt to the other side my man. That's where the danger is." The soldier responded in a distinct southern drawl, "Don't you suppose I know which side them Yanks are on?" Chamberlain put on his most authentic southern drawl and said, "Dig away then but keep a sharp lookout," and backed away into the darkness.

The next major battle of the two armies, Chancellorsville, did not involve any action by Chamberlain or the 20th Maine. Then, as you know, after Chancellorsville, which was a total Confederate victory, Lee resolved to take the offensive and move into Pennsylvania to invite a major battle and by the Confederate victory he expected, to encourage a peace movement among the states of the north. Over the next eight weeks both armies moved in parallel lines, the Union army some 50 miles to the east of and behind the Confederates, into the rolling hills of southern Pennsylvania. On July 1, the advanced brigades of each army met a few miles northwest of Gettysburg and, after the first day of fighting, the Confederates had pushed the Union forces southeast into and through the town of Gettysburg and onto the hill behind it known as Cemetery Ridge. During the day of July 2 most of the Confederate forces began pressuring the length of the Union line which itself was being steadily reinforced as more troops moved up from Washington.

All of you are very likely familiar with the lines of battle of the opposing armies at Gettysburg. The position of the Union forces has often been characterized as a fishhook with the hook and barb at the northern and northeastern end of the position which was the Union's right flank, the length of the hook running north to south along Cemetery Ridge and the eye of the hook at the southerly extremity of Cemetery Ridge on a rocky hill called Little Round Top, the extreme left flank of the Union forces. As Chamberlain's regiment moved on to Cemetery Ridge in the late afternoon of July 2, the Confederates, under Longstreet, were attacking in echelon against the Union left. It happened that Major General Daniel Sickles of New York had moved his two divisions, which were then on the extreme left of the Union line, forward from Cemetery Ridge across a little stream and on to higher ground about 1/2 mile west of the Union's main line of battle. Sickles' position encompassed a peach orchard and, on his left, a rocky ravine called the Devil's Den. His advance from Cemetery Ridge had also created an untenable bulge in the Union line. Even as Chamberlain arrived, the Confederate troops could be seen from Cemetery Ridge advancing around Sickles' left and effectively outflanking the entire Union army. Regiments from Maine, New York and Pennsylvania were rushed along Cemetery Ridge to meet the on-rushing Confederate attack in the south. As Chamberlain made his way toward the south end of the Union line, he met Colonel Strong Vincent. Vincent pointed out to Chamberlain the ground that the 20th Maine was to defend, telling him that it marked the extreme left of the entire Union line. Vincent explained that a "desperate attack" was expected at any moment to turn the Federal flank. Chamberlain recalled Vincent saying to him as he reached Little Round Top: "I place you here. This is the left of the Union line. You understand. You are to hold this ground at all hazards." The military consequences of failing to hold and to have the left flank of the entire army turned meant that the Confederates could seize the heights, possibly gain the rear of the whole Union position, take or destroy the supplies and ammunition behind the lines, and effectively roll up the Federal line to the north.

Chamberlain's forces on the southern portion of Little Round Top comprised 308 men and two dozen officers. This was all that was left of a regiment that had been nearly 1,000 strong only 10 months before. In addition, the day before he had been put in charge of a group of approximately 100 men from another Maine regiment that had mutinied two weeks earlier in Virginia. All but a handful of those prisoners joined the 20th Maine in the defense of Little Round Top.

Within minutes after Chamberlain arrived, the battle was joined. The estimate was that had Chamberlain reached Little Round Top fifteen minutes later, Longstreet's regiments would have been in possession of it. In fact, after the war, Longstreet cut the margin even closer. "I was three minutes late in occupying Little Round Top," he told a group of Union Gettysburg veterans. "If I had got there first you would have had as much trouble in getting rid of me as I did in trying to get rid of you."

Even with the mutineers, Chamberlain's regiment was less than half strength, and it was opposed by at least one Alabama brigade with more than three times the number of men. Between 5:30 p.m. and twilight, the Confederates mounted four charges against Chamberlain's position. After the third charge the 20th Maine's casualty count passed 100. Chamberlain had already received a bullet wound to his right instep and a minie ball had struck the scabbard of his sword, causing a flesh wound. The fourth charge by the Alabamians ran out of momentum just as the 20th Maine's fire began to fall off, due to lack of ammunition. Hand-to-hand action was present all along the line. By now both sides had launched thousands of minie balls at each other. Fighting had been going on for nearly two hours. The fire had been so intense that most of the trees in the area were shredded at a height of 5 or 6 feet, and many of them were literally sawed in half by the gunfire.

Just before dark the Confederates re-massed and prepared for a fifth charge. By now most of the 20th Maine had used up all of their issue of 60 rounds, and only a handful of men were able to return fire. More than 150 of his men were dead or wounded. As the Confederates began moving up through the trees again, Chamberlain recalled later that his only thought at that time was Col. Vincent's order to hold the ground at all hazards. Chamberlain shouted to his men, "I am about to order a charge. Beginning on the left we will make a great right wheel." He stepped to the regimental colors and shouted up and down the line, "Fix bayonets." Then he turned, stepped off the rock and led his men down through the trees and into the advancing Confederates. The charge turned the Confederate line. At the base of Little Round Top an Alabama officer aimed a revolver at Chamberlain's face. The weapon misfired. With his free hand he offered Chamberlain his sword. Scores of Confederates surrendered to the 20th Maine and the Union's left flank was secure. For his action at Little Round Top Chamberlain was awarded the Congressional Medal of Honor.

For the next ten months following the Battle of Gettysburg, Chamberlain was essentially inactive, first with a 20-day furlough in August 1863 and then, after contracting malaria in the autumn, for nearly six months on extended sick leave. Unable to sit still at home and eager to be back in the army, Chamberlain re-joined his regiment in May, 1864. Following Gettysburg, Chamberlain's immediate superiors had recommended him for Brigadier General. Although the appointment did not come through from Lincoln, Chamberlain, although only a colonel, was assigned by General Warren to take charge of five Pennsylvania regiments consolidated to form a new brigade, the 1st Brigade of the Fifth Army Corps. The Union Army had by now reached the Petersburg defenses. On June 18 Chamberlain was requested to lead a major assault by the better part of two Union divisions against a heavily fortified part of those defenses. As was his custom, Chamberlain stepped to the front of his leading brigade, drew his sword, exhorted his men to follow him and turned and began a run towards the Confederate line. It was during this charge, as he half turned to encourage his men onward that Chamberlain was hit by a bullet. The bullet entered below his right hip. Unable to run further he attempted to stand by leaning on his sword, but within a few seconds sank to the ground and was fortunately carried off by two subalterns who moved him out of the direct line of fire. Carried to a field hospital, the field surgeons determined that the minie ball had entered Chamberlain's right hip, struck and splintered his hip bone, and in its passage cut the bladder and urethra. They opened Chamberlain's abdomen, cleaned out the bits of cloth and bone, extracted the ball which had lodged on the inside of Chamberlain's left hip, made some crude repairs by sewing up the bladder and reconnecting as best they could the severed urethra, and they closed him up. As was so often the case with those who were "gut shot" during the war, Chamberlain was given no chance of survival, and with the inevitable worsening of his condition, his obituary was released to the newspapers. He dictated his last letter to his wife and family.

Much of the credit for Chamberlain's survival went to the surgeons who attended him and who had known him well previously. They had a personal stake in his recovery. Still, Civil War medical techniques were primitive, even in the most competent hands. There is a story, for example, that when Oliver Wendell Holmes, Jr. took a minie ball in the heel at Chancellorsville, surgeons shoved a raw carrot into the wound to keep it open and allow it to drain.

Expected to die, he was nevertheless promoted to Brigadier General in the field, the first such promotion made by Grant during the war. By some miracle he clung to life and was removed to the Naval Academy Hospital at Annapolis. He continued to improve and by September was well enough to be discharged from the Hospital and sent to his home in Brunswick for convalescence. Bowdoin had requested him to return to teaching when he was fully recovered, but by the end of October he felt himself drawn again to the war, and on October 29, 1864 still racked with abdominal pain, he boarded a train to rejoin the Army of the Potomac. But he was not fully recovered, and after participating in an attack on the Weldon Railroad which involved a five-day raid in rain, snow and freezing weather, he was sent to Philadelphia in early January for further abdominal surgery and convalescence again in Maine. He was encouraged to remain and continue teaching at Bowdoin, and he was offered a position as Collector of Customs. But by now he was obsessed with returning to the war (which he characterized as "most congenial to my temperament") and in late winter he rejoined his brigade over the vigorous opposition of his family and friends.

He returned in time to play a pivotal role in the Battle of Five

Forks in late March, 1865, the last major battle of the War. Leading his Brigade in a charge against a full Confederate division and mounted on his favorite horse, Charlemagne, Chamberlain was struck in the chest just below the heart and he collapsed onto the horse's neck, feeling blood there. It so happened that the ball that had struck him had first passed through Charlemagne's neck and then glanced off the frame of a hand mirror in Chamberlain's breast pocket. From there it followed the curve of his ribs under his skin and exited out of his back and the back seam of his coat, speeding on to its next target, the pistol case of the aide riding along beside him, a subaltern named Theodore Vogel. The impact of the minie ball knocked Vogel right out of his saddle. Although Chamberlain was stunned from the blow and was thought by some of his lieutenants to be dying, he mounted a stray horse, rode to the front of the line that was then collapsing from a Confederate countercharge, stemmed the retreat and led a fresh charge that resulted in a full Confederate retreat. Over the course of the battle his Brigade captured more than 1,000 of George Pickett's veteran soldiers.

His conduct at Five Forks earned him a brevet promotion to Major General for conspicuous gallantry, and it brought him the enthusiastic attention and support in these final days of the war of General Phil Sheridan whose cavalry had been crucial to the Union victory at Five Forks. Moving with Sheridan following the Battle of Five Forks, he performed an all night march with his Brigade to keep up with Sheridan's cavalry in pursuit of what remained of Lee's army. As dawn broke on April 9, 1865 Chamberlain stood with the Fifth Corps on a hill gazing down on the Appomattox River Valley, the little community of Appomattox Courthouse, and 20,000 men of the Army of Northern Virginia massed below. As his brigade advanced down the slope Chamberlain was approached by a southern cavalry officer carrying a white flag. Lee and Grant were to meet to discuss surrender. At about midnight that same evening, Chamberlain was informed that he had been chosen by Grant to command the formal surrender ceremony, although he was never given the reason why he was selected.

At first light on April 12, 1865, a chill, gray morning in the Valley of the Appomattox, Chamberlain formed his division along the main road leading up from the Appomattox River into the town. On the far side of the valley the Confederates were striking and folding their tents and forming ranks for this last ceremony - that of formally laying down their arms. The ragged column of Confederates was led by General John Gordon on horseback. As the column approached, Chamberlain called out the order for his entire division to carry arms - the marching salute. The entire 1st Division snapped to attention, from right to left all along the line, from the river bank to the center of the little town where the weapons were to be laid down. General Gordon was clearly taken by surprise, as were his soldiers. With a flourish of his sword, he bowed low over his horse and touched his sword tip to his boot. Then he turned to his men and gave an order which was repeated back through the ranks. The Confederate flag dipped and the surrendering soldiers returned the salute. Chamberlain spoke of this later as "honor answering honor." The nobility of his gesture was much publicized throughout the North where it was recognized as a first generous step in the process of healing the wounds between north and south. And General Gordon called Chamberlain "one of the knightliest soldiers of the Federal army."

Chamberlain was as thorough and complete a hero as had emerged from the war. Honored by his superior officers and by the President of the United States, he returned home to the accolades of his countrymen in the State of Maine. But within a year he found his return to a more cloistered life, even one that found him at the center of Maine's society, to be disappointing.

Searching for something akin to the activity and excitement of the war, Chamberlain turned to politics shortly after he returned, and in September 1866 he was elected governor of Maine by a landslide. Chamberlain won re-election as governor in 1867, 1868 and 1869. Following his fourth term as governor, his friends promoted his candidacy for the U.S. Senate, although Chamberlain's views were then at odds with those of the mainstream Republicans in Maine at the time, and the Republican caucus chose another candidate. Chamberlain believed in a generous reconstruction policy, and he favored the complete withdrawal of Federal troops from the south, whereas the majority of Maine Republicans at the time were in a more punitive frame of mind.

Chamberlain completed his fourth and last term as governor in 1870 and returned to Brunswick where the Trustees of Bowdoin offered him the presidency of the College in the summer of 1871, with an annual salary of $2,600. He accepted at once.

But if Chamberlain had enjoyed public celebrity and success during and after the war, the same could not be said for his personal life. His relationship with Fanny, strained by the war and politics and his clear preference for them over a family life, began to disintegrate. She grew resentful. She refused to join him in Augusta during his years as governor. Chamberlain was himself often moody and morose. His abdominal injuries caused him great pain much of the time, and apparently rendered him impotent or nearly so. There were rumors that Chamberlain actually physically abused his wife, although these were only rumors and seemed to be the product of Chamberlain's political adversaries, of which he had many. In all events, Chamberlain and Fanny never separated and over the years were reconciled and seemed generally happy together. Chamberlain rebuilt his home across from the Bowdoin campus, turning it into a 20-room Victorian mansion. There was a great deal of space for entertaining, and Grant, Sherman, Sheridan and McClellan all came to visit and stay. Even Longfellow came and spent several days with Chamberlain in 1875 on the 50th anniversary of his graduation from Bowdoin.

Chamberlain served as President of the College for 12 years, from 1871 to 1883. His tenure as President was known for a comprehensive overhaul of the curriculum. Chamberlain's predecessor had added some science courses to the curriculum, and had encouraged the separation of science and theology, although Paley's Evidence of Christianity remained on the list of required reading for all seniors. Chamberlain carried the reforms a stage further by establishing a scientific department that awarded a Bachelor of Science degree. He expanded the study of languages, including French and German, and gradually introduced the system of elective courses, moving away from a rigid curriculum over each of the four years. During his tenure women were permitted to audit some courses. He encouraged a greater study of music and the arts.

Chamberlain also required military training, a clear carry-over from his wartime experience. He believed it was a means, short of war, to instill the qualities of discipline and courage that he so admired. Interestingly, at the end of his tenure as President, he argued that football, then in its embryonic stage, provided a means to the same end. It was his insistence upon strict military training that caused the one serious crisis he faced as President. Chamberlain's implementation of military training culminated in an order requiring each student to purchase a $6 uniform and, as a result, an uprising known as the "drill rebellion" occurred. It pitted Chamberlain against nearly all the students at the school. Chamberlain reacted by suspending the entire junior, sophomore, and freshman classes, and sending all of those students home. The parents and many of the faculty complained of his heavy-handed tactics, and Chamberlain found that he had painted himself into a corner. Finally and mercifully the Trustees intervened, offering a compromise that allowed the suspended students to return. It was a defeat for Chamberlain, and one which brought him no credit.

Always eager for action, Chamberlain had his last hurrah as a soldier in 1879 when the citizens of Maine were violently split between Greenbackers and those who wanted to retain the gold and silver standards. When state elections, including that of the governor, were challenged and overturned, civil order broke down and the state was on the verge of civil war. Chamberlain was summoned by powerful state legislators to restore order. Effectively declaring martial law, Chamberlain padlocked the governor's office and restricted access to the legislature. He kept troops stationed at various key points throughout the state in the event of mob activity. In the meantime he asked the Supreme Court of Maine to issue a definitive ruling on the legitimacy of the elections. In the meantime the democrats chose a governor, but Chamberlain, then effectively wielding autocratic power, declined to recognize him. The Republicans then offered the governorship to Chamberlain, and he rejected this as well. Over a period of weeks there were plots and counter-plots and threats to kidnap or kill Chamberlain. Finally, with a definitive decision by the Maine Supreme Court and the election by the Legislature of a new governor, Chamberlain recognized the election as valid, disbanded his militia, and peace was effectively restored.

In 1883 Chamberlain stepped down as President of Bowdoin. He had taught every subject at the college except mathematics. He could not know it but 30 years of life were still left to him.

Somewhat at loose ends, he dabbled in different ventures, including real estate investments in Florida. After several trips to the Ocala area he formed and became President of the Florida West Coast Improvement Company. His objective was to make money through land speculation, as well as find a pleasant winter climate for himself and Fanny as they grew older. But the Florida venture went nowhere, and he lost a fair amount of money in the process. The golden days of Florida land speculation were another 75 years away.

While he still had money, he and Fanny would spend part of each winter in New York City, taking rooms on 57th Street near Central Park. When his health permitted, which was less and less as his war wound flared up more often, they joined the winter-time social rounds and had an active social life. Chamberlain also had a Wall Street office that he used for his various business activities, including his Florida venture.

By the mid 1890's, however, with Chamberlain in his 60's, he found himself down on his luck financially. He sought various appointed governmental positions, sinecures really, but was unsuccessful. When the Customs Collectorship of Portland, Maine became vacant in 1899, Chamberlain and his friends campaigned hard for that position, considered the State's most lucrative patronage office. "He is poor and afflicted and his wife is totally blind," a former Maine Supreme Court Justice wrote in a letter recommending Chamberlain for the opening. One of Chamberlain's old 20th Maine comrades, Ellis Spear, wrote to the congressman representing the Portland area that such an appointment would "save Chamberlain's last days from the distress of penury." Nonetheless, the office went to another individual, but in March, 1900 a consolation prize, the Port Surveyorship, came to Chamberlain. It carried less responsibility and prestige, although the salary of $4,500 a year was generous at the time.

In the early years of the new century, there was a resurgence of interest in the Civil War, and Chamberlain found that there was a vigorous demand for him to speak and write. He took to the lecture circuit throughout the northeast and was quite successful, and it provided a helpful supplement to his income. He had remained handsome, and his white hair and mustache and his proud bearing presented a paradigm of the old soldier/hero. Although his speeches were somewhat flowery and grandiloquent, his voice remained strong and stammer-free, and his words often brought many in his audience, especially the older Union veterans, to tears. He wrote numerous magazine articles and a book that was posthumously published as "*The Passing of the Armies*."

From his earliest days, indeed beginning as early as 1864, Chamberlain made tours of the battlefields, and he frequently took part in the Gettysburg rites. He undertook a final journey to Gettysburg in May, 1913. Abner Shaw, the 20th Maine surgeon who had helped save his life at Petersburg, accompanied him. One evening, an hour or so before sunset, they trudged together up the overgrown slope of Little Round Top and sat down among the crags. Nearly 50 years earlier, at about the same time of day, Chamberlain had ordered the bayonet charge that had in all likelihood preserved the victory at Gettysburg for the Union forces.

Unlike Sherman who claimed that "war is hell," Chamberlain saw it very differently. He said at one point:

Fighting and destruction are terrible; but are sometimes agencies of heavenly rather than hellish powers. In privations and sufferings endured as well as in the strenuous action of battle, some of the highest qualities of manhood are called forth - courage, self-command, sacrifice itself for the sake of something held higher . . . fortitude, patience, warmth of comradeship, and in the darkest hours tenderness and caring for the wounded and stricken . . . .

For Chamberlain war was romantic, heroic, a test of character. It is no small irony that World War I commenced less than a year after his death, resulting after 1918 in a profound disgust for war on the part of the victors that would have been wholly out of step with Chamberlain's belief in the nobility of combat.

Joshua Lawrence Chamberlain died of complications of his Petersburg wound shortly before noon on a bitterly cold February 24, 1914 in Portland, Maine. His funeral was a celebration of Maine's greatest hero.

BIBLIOGRAPHY

Golay, Michael, *To Gettysburg and Beyond*; Crown Publishers, Inc. 1994.

Shaara, Michael, *The Killer Angels*; Ballantine Books 1974.

Trulock, Alice *Rains, In the Hands of Providence*; The University of North Carolina Press 1992.

The Round Table Presentation | June 4, 1998

Capturing Time
John Harrison And His Extraordinary Clocks

In October 1707 Admiral Sir Clowdisley Shovell led four British Men-O-War home from Gibraltar after victories over the French forces in the Mediterranean. Beset by fog for most of the return trip and afraid that his ships might founder on British coastal rocks, the Admiral summoned his navigators. It was their consensus that the English fleet was safely west of the coast of England. But on the foggy night of October 22, 1707, the four ships discovered to their horror that they had misgauged their longitude and were bearing down on the jagged coast of the Scilly Isles, just off Land's End. All the ships and over 2,000 men were lost.

The loss of that squadron, so many ships, men and England's maritime reputation all at one time, was a principal factor in the enactment of the Longitude Act of 1714 by which Parliament offered £20,000 (the equivalent of many millions of dollars today) for a method to determine longitude at sea to an accuracy of ½ a degree.

Since 1 degree of longitude spans 60 nautical miles (the equivalent of 68 geographical miles) over the surface of the globe at the equator, even ½ a degree error, or 30 miles, translates into a large distance. The fact that Parliament was willing to award such huge sums for methods that were still wide of the mark suggests the crude and, one might argue, desperate state of navigation as late as the 18th century.

The Longitude Act established the Board of Longitude, consisting of distinguished scientists, naval officers and government officials who were given sole discretion over distribution of the prize money. It included the Astronomer Royal as well as the President of the Royal Society, and it was empowered to distribute lesser amounts to help needy inventors bring promising ideas forward. Parliament was not the first to seek a solution to the longitude problem. Both King Philip of Spain and King Louis XIV of France had set up prizes and awards for useful solutions. But the sheer magnitude of the British prize, based as it was on the dependence of the British people upon the success of their mariners, attracted the attention of astronomers and inventors throughout Europe.

Just what was this problem about longitude? Lines of latitude are parallels, with the zero degree parallel being fixed at the equator as early as 150 B.C. by the cartographer and astronomer Ptolemy. From the earliest times any sailor worth his salt could find his latitude well enough by the height of the sun or certain known guide stars above the horizon. Any capable sailor, including Christopher Columbus, could sail the parallel as he did on his 1492 journey. The measurement of longitude meridians, by comparison, has no such easy reference as the angle of the sun or of certain stars. Indeed, longitude is immutably tied to the concept of time, the rotation of the earth. To learn one's longitude at sea, one needs to know what time it is aboard ship (primarily accomplished by establishing local noon by taking angular sightings of the sun), and also the time at the home port or another place of known longitude at the very same moment. The difference in the two clock times allows the navigator to measure a geographical distance. Since the earth takes 24 hours to complete one full revolution of 360 degrees, one hour marks 1/24 of a spin, or 15 degrees. And so each hour's time difference between the ship and the starting point marks a progress of 15 degrees of longitude to the east or west. Every clear day at sea, after 1770 and before satellite navigation, when a navigator reset his ship's clock to local noon when the sun reached its highest point in the sky, and then consulted the home-port clock, each hour's discrepancy between the two times translated into another 15 degrees of longitude.

Of course those same 15 degrees of longitude also correspond to a distance traveled. At the equator 15 degrees stretch approximately 1,000 miles. North or south of that line, however, the mileage value of each degree decreases. One degree of longitude always equals four minutes of time the world over, but in terms of distance, 1 degree shrinks from 60 nautical miles or 68 land miles at the equator to almost nothing as one nears the poles.

At the time of the passage of the Longitude Act, the clockmaker's art was still far from developed. Consequently, for want of a means of determining longitude, every sea captain, including the great captains of the age of exploration, daGama, Magellan, Drake, had to find their way about the ocean's surface by dead reckoning, a combination of knowing one's latitude, but then having to determine longitude by the estimated speed of the ship and the sideways movement that could be attributed to sea currents or the quartering winds.

The hero of this story is an English clock maker named John Harrison. With no formal education or apprenticeship to any watchmaker, Harrison was able to construct a series of virtually friction-free clocks that were made from materials that were impervious to rust, and that kept their moving parts perfectly balanced in relation to one another, regardless of how much they were pitched or tossed by the seas. He did away with the pendulum, and he combined different metals inside his works in such a way that when one component expanded or contracted with changes in temperatures, the other counteracted the change and kept the clock's rate constant. His invention was a clock that could carry the true time of the ship's home port to any remote corner of the world and allow a sea captain to know his longitude to within a matter of several miles. By the end of Harrison's life he had advanced the technology of clock making, and indeed the technology of many mechanical processes, to such a degree of perfection and accuracy that leading scientists of the day could scarce believe it. But before we describe Harrison's achievements, it may be worthwhile to look briefly at the development of clock making over the centuries.

We know that the measurement of the year and the month was performed almost as far back as we have records, but the shorter units of time remained vague and did not play a large part in the human experience until the Egyptians who perfected the use of the sun dial at least as early as 1500 B.C. in the reign of Thutmose III. Interestingly, these early sundials divided the day into 12 equal parts.

The Greeks and the Romans made numerous improvements in the sundial, but sought ways to measure the dark hours. The most widely used of these methods was the water clock, or klepsydra, meaning water-thief, and water clocks became a popular and widely-used method of telling time indoors. Legal arguments in the courts of Athens and Rome were measured by water clocks.

In medieval Europe sand glasses became popular once the glass maker's art had been mastered to create a properly calibrated and sealed container. Sand would flow in climates where water would freeze, and the self-contained nature of sand glasses made them more portable than water clocks.

But the first methods for the mechanical measurement of time were developed not by merchants or craftsmen. Instead they originated in monasteries whose monks needed to know the time for their appointed prayers. These first clocks were designed not to show the time, but to sound it. Monastic clocks were weight-driven machines which struck a bell after a measured interval. They announced the canonical hours, the times of day prescribed by the church for devotion. Indeed the word clock bears the mark of its origins. It is derived from the Middle Dutch word glocke, which means bell. So in the beginning, these kinds of mechanical devices were not considered to be clocks unless they rang bells.

Bell clocks worked by the force of falling weights. What made these machines workable was the device that prevented the free fall of the weights and interrupted their drop at regular intervals. This simple device which has remained relatively uncelebrated in history, was called an escapement since it was a way of regulating the escape of the gravitational pull of the weights. The escapement was so designed that it would alternately check and then release the force of the weight on the moving machinery of the clock. This was the basic invention that has made all modern clocks possible. The regular downward pull of the falling weights was translated into the interrupted, regular movement of the clock's machinery.

It is not known exactly when these canonical clocks were first invented and used, but we do know that it was not until the 14th century that these clocks began to have wider use outside the monasteries. It was also at this time that the hour became generally accepted in Europe as one of 24 equal parts of a day. Why 24 hours? History is not clear. The Egyptians apparently divided their day into 24 hours, and this early division seems to have carried down the centuries. It was also in the 14th century that the notion of minutes and seconds was developed. This seems to have come originally from the Babylonians who had adopted a sexagesimal system. The notion of second is derived from the second subdivision on the basis of 60. Since Ptolemy and other early astronomers used a 60-unit system for subdividing the circle, it is not difficult to conclude that this same approach was adopted for the keeping of time.

By the 14th century, mechanical masters around Europe began developing great mechanical clocks. Sometime during this period clocks added dials so that the measurement of time became a visual experience as well. A magnificent clock made about 1350 for the Cathedral of Strasbourg served the public both as a calendar and an aid to astrology. Using an enormous number of gears and cogged wheels, it performed a variety show as it tolled the hours. On its numerous dials it recorded the hour and minute, the times of the setting and rising of the sun, the day of the month and month of the year, the fixed feasts of the Church, the length of daylight for each day, the solar and lunar cycles, the annual movement of the sun and moon, and the movements of the five known planets.

Many of you may know the story how in 1583 Galileo, then a 19-year old attending prayers in the baptistery of the Cathedral of Pisa, was distracted by the swinging of the altar lamp. No matter how wide the arc of the lamp, it seemed that the time it took the lamp to move from one end of the swing to the other was always the same. Galileo timed the intervals of the swing by his own pulse, since he had no other device for time measurement. Galileo had discovered what has come to be known as the isochronism, or equal time of the pendulum: that is that the time of a pendulum's swing does not vary with the width of the arc, but only with the length of the pendulum itself. This simple discovery revolutionized clock making. It made clocks portable, and by 1650 the average error of the best time pieces was reduced from 15 minutes to less than one minute per day.

Now Galileo was no sailor, but even he had heard of the longitude problem, as had every natural philosopher of his day. The pendulum clock, a portable clock that could be carried aboard a ship, was a possible answer. While Galileo himself never developed a pendulum clock, a brilliant young Dutch mathematician and astronomer, Christiaan Huygens, devised a portable pendulum clock at the age of 27 in 1656. But though his pendulum clock worked extremely well on land, the pendulum could not keep accurate time on a surging, rolling ship. Huygens' efforts made it clear that the longitude problem could not be won by a clock driven by weights and pendulums.

The alternative was a spring which could propel the timepiece as the spring unwound. The Italian architect Brunelleschi created a spring-driven clock as early as 1410, and within a century clocksmiths were making relatively small clocks driven by springs rather than pendulums. But the spring had its own problems. While a weight descending exerts the same force whether at the beginning or at the end of its descent, an uncoiling spring will exert a decreasing force as it unwinds. An ingenious solution to this problem was the fusee, a conical spool so designed that as a spring unwound the spool exerted an increasing force on the machine. Throughout the 17th century many inventors tried to perfect the spring-drive clock, including Huygens in Holland and Robert Hooke in England. But none of them could create a clock that would keep accurate time on a sea voyage. Most would lose minutes each day, making the effort to tell longitude from a clock more of a menace than a help.

At the time the Longitude Act was passed in 1714, the scientific world was of two minds regarding the determination of longitude at sea. One possible solution was, of course, the use of a chronometer to measure the time between one's home port and a position at sea. The other was the application of astronomy, particularly the movement of the moon through the heavens. It had long been thought that longitude might well be determined by reference to the location of the moon to certain fixed stars in the sky. Following the passage of the Longitude Act, astronomers in England at the observatory in Greenwich began a massive project to map the stars and the progress of the moon through the night heavens. It is of interest that some of the scientists who sought to determine longitude were equally facile in both the areas of mechanics and astronomy. Huygens, the maker of the pendulum clock, was also a gifted astronomer who first discovered the rings of Saturn. Galileo, the discoverer of the pendulum, had also invented the telescope. He had found that four large satellites of Jupiter orbited that planet with such regularity that his observations could be used to create tables to calculate longitude. Indeed, after about 1650, Galileo's method for using the moons of Jupiter to find longitude became generally accepted on land, and cartographers used Galileo's technique to redraw the map of Europe. But seeing the eclipses of the Jovian satellites on the deck of a pitching and rolling ship was a technological problem that no one could solve.

Sir Isaac Newton, who was then head of the Royal Society and also a member of the Board of Longitude, after reviewing both astronomical and chronometric proposals, concluded that any hope of settling the longitude matter lay in the stars. The lunar distance method had been proposed several times, and had gained credence and support as the science of astronomy improved. In a letter from Newton to the Secretary of the Admiralty in 1721, he said:

> "*A good watch may serve to keep a reckoning at sea for some days and to know the time of a celestial observation; and for this end a good jewel watch may suffice till a better sort of watch can be found out. But when the longitude at sea is once lost it cannot be found again by any watch.*"

Into this set of circumstances stepped John "Longitude" Harrison. Born on March 24, 1693 in Yorkshire, the eldest son of a carpenter, he received no formal education, but was immensely curious and learned to read and write on his own. But not until he was 19 was he able to get his hands on a textbook on natural philosophy. From that he moved to Newton's *Principia Mathematica*, and his many notes taken relating to his studies reflect his full understanding of what he read.

Harrison completed his first pendulum clock in 1713 before he was 20. He had had no experience as a watch maker's apprentice, and it is not clear how and why he took on the project. What gives his first clock even greater uniqueness is that it is constructed almost entirely of wood, with oak wheels and axles moving small pieces of brass and steel. The clock still exists at the Worshipful Company of Clock Makers Museum in Guildhall in London. The wooden teeth of the wheels have never snapped off, but have defied destruction by the unique design which allows them to draw strength from the grain pattern by which they were cut.

Then, sometime about 1720, after Harrison had apparently acquired a reputation as a clockmaker, he was hired to build a tower clock at a manor house in Brocklesby Park. Harrison completed the clock in 1722, and it still tells time in Brocklesby Park. It has been running continuously for more than 270 years, except for a brief period in 1884 when it was stopped for refurbishing. As with so many of his other creations, this one was exemplary. For one thing, it runs without oil. In fact this clock never needs lubrication because the parts that would normally call for it were carved out of lignum vitae, a tropical hardwood that exudes its own lubricant. Harrison avoided the use of any metal in the clockwork mechanics for fear that it would rust in the damp conditions. Whenever he needed metal he installed parts made of brass. Like his first pendulum clock, the toothed gears are made of oak. His knowledge of wood, its strengths and resiliency allowed him to fabricate a clock that has virtually friction-free gearing. And it may be that Harrison, who undoubtedly knew of the longitude prize at this point, may have already been thinking about a clock that would not need oil for its gears. A clock without oil, which till then was totally unheard of, could stand a much better chance of keeping time at sea because lubricants got thinner or thicker as temperatures dipped or soared over the course of a voyage, making the clock run faster or slower as a result.

In the 1720's Harrison built additional pendulum clocks of greater and greater precision. He began creating pendulums out of alternating strips of two different metals. He realized that pendulums expanded with heat, so that they grew longer and ticked out time more slowly in hot weather. In cold weather they contracted and this speeded up the seconds, throwing the clock's rate off in the opposite direction. What Harrison combined were two metals - brass and steel - whose characteristics counteracted each other, so that the pendulum never went too fast or too slow.

Harrison's meticulous notes tell us that he tested these clocks over the course of many months by noting the time they kept in comparison with the movement of certain stars across a fixed area of the heavens. In these tests, Harrison's clocks never erred more than a single second in a whole month. In comparison, the most expensive clocks being produced anywhere in the world at that time usually lost or gained about one minute every day.

By the late 1720's, Harrison was focused on the longitude prize and the special challenge of time keeping aboard ship. He had already found a way around the problem of lubricants, he had created a mechanism that was virtually friction-free, and his clocks kept time with unparalleled precision. But if he were going to make a sea clock, he would have to jettison the pendulum. No pendulum could withstand the movement of a rolling ocean. What Harrison began to envision in lieu of a pendulum was a set of see-saws motivated by springs, self-contained and counter-balanced to withstand the most extreme motion. It took him nearly four years to work out the design for his sea clock, and when he did he set off for London in 1730 to submit his proposal to the Board of Longitude. Harrison went first to Dr. Edmond Halley, the Astronomer Royal, who was busy mapping the heavens at the royal observatory at Greenwich. Halley, although an astronomer, seems to have been open to a mechanical answer to the longitude problem. Halley realized that the Board of Longitude, made up as it was of admirals and astronomers who favored astronomy, may be hostile to a clockwork solution, so he referred Harrison to a well-known watch maker named George Graham. Graham was fascinated by Harrison's proposal and sent him on his way with a generous loan.

For the next five years Harrison worked building his first sea clock, which has come to be called *Harrison's Number One*, or H-1 for short. Weighing 75 pounds and looking like a Rube Goldberg contraption with rods and balances sticking out at weird angles, together with four strange looking dials, H-1 was an absolute marvel.

Harrison presented it to George Graham in 1735 who was so delighted with it that he in turn displayed it before the members of the Royal Society who were equally impressed and recommended it to be tested for the prize.

After a year of waiting, the clock was put aboard a ship to sail to Lisbon. On its return voyage it proved to be absolutely accurate in predicting the location of various islands and navigational points. In consequence, upon his return to London, the Board of Longitude convened for the very first time 21 years after it was created to sit in judgment on his clock. The Board was disposed to give the clock a trial to the West Indies and back (which round-trip thereafter became the acid test for any chronometer that might claim the prize). But it was Harrison himself who demurred. He said that the clock showed some defects that he wanted to correct. With another two years' work if the Board could advance him some funds, he thought he could produce an even better timekeeper.

Harrison's No. 1 became quite a famous curiosity in London at the time. George Graham had it on loan from Harrison and kept it exhibited in his shop where people came from England and from the continent to see it and marvel at the ingenuity involved. The English artist, William Hogarth, who had an obsession with time and time-keeping and who had actually begun his career as a watch case engraver, described Harrison's No. 1 as "one of the most exquisite movements ever made."

Incidentally, H-1 continues to run with daily winding at the National Maritime Museum in Greenwich.

In 1741 Harrison completed his second great sea clock, H-2, but by the time he presented the clock to the Board of Longitude, he was already disgusted with it. All he really wanted, he said, was their blessing to go home and try again. As a result, H-2 never went to sea. It was, however, tested by the Royal Society, who subjected it to heating, cooling and agitation for many hours together, with greater violence than it could receive from the motion of a ship in a storm. The Royal Society concluded that:

> *"The result of these experiments is this: that (as far as can be determined without making a voyage to sea) the motion is sufficiently regular and exact, for finding the Longitude of a Ship within the nearest limits proposed by Parliament and probably much nearer."*

What Harrison did after H-2 was to exhibit the characteristics of a perfectionist. He retreated into his workshop, and barely anything was heard from him during the next 19 years that he devoted to the completion of Harrison's No. 3. He did request and receive occasional stipends from the Board of Longitude as he worked through the problems of replacing his bar-shaped balances with circular balance wheels.

With Harrison's disappearance from the scene, the focus

turned to the exponents of the lunar distance method. This group, consisting of the Astronomer Royal and his staff, admirals and sea captains, had gained ascendancy because of the significant number of observations that had by then been made to allow a fairly complete mapping of the heavens, at least as far as the major stars were concerned. The lunar method relied on three pillars for support. First, it was based upon the motion of the moon vis-à-vis certain fixed stars in the sky. Second, it had gained momentum after the invention of the sextant which allowed angular distances between the moon and fixed stars to be calculated with greater ease. Third, it required reference to tables to allow for the calculation of longitude. Much of the problem with the lunar method lay in the tables themselves. Because the moon's elliptical path wanders through the night sky in an extremely complex rhythm, extensive tables had to be created to account for the moon's varied orbit. But there were other problems, too. For approximately five days each month, either the moon's nearness to the horizon, or its new moon phase, would make this method unusable. Light refraction, lunar parallax, and a thorough understanding of how to apply the tables made finding longitude by the lunar method one of the most difficult sets of calculations one could imagine. Nonetheless, by the late 1750's the lunar technique finally looked practicable.

In the meantime, John Harrison labored on clock number 3. Sequestered for a great part of the time, he did emerge in 1749 to accept the Copley Gold Medal for his efforts in creating H-1 and H-2. Later recipients of the Copley Medal include Captain James Cook, Benjamin Franklin, Ernest Rutherford and Albert Einstein.

Harrison's No. 3, the lightest of his sea clocks, weighed only 60 pounds. Containing 753 separate parts, it also included numerous innovations including something referred to as a bimetallic strip, consisting of brass and steel melded together. The bimetallic strip immediately and automatically compensates for any changes in temperature, and even today is used in thermostats and temperature control devices. He also developed caged ball bearings for use inside H-3. This invention survives today in practically every machine with moving parts.

Harrison substantially completed H-3 in 1757 and presented it to the Board of Longitude in 1759. He was quite satisfied with it and spoke of how much he had learned in producing it: "[The] transactions I had with my third machine [were] so very weighty [and] so highly useful [that these discoveries] were worth all the time and money it cost viz my curious third machine."

But H-3, while a mechanical and technological marvel, was not destined to be the piece de resistance of Harrison's creations. That distinction belonged to Harrison's No. 4 which was not a clock at all but a somewhat oversized pocket watch.

Following the brass and steel behemoths of H-1, H-2 and H-3, H-4 is only 5 inches in diameter and weighs only 3 pounds. If you looked at Harrison's clocks as representing a logical, linear progression of his inventiveness, H-4 is a complete non-sequitur. As one writer describes it, "Given what Harrison had already produced, H-4 is as surprising as a rabbit pulled out of a hat." How he was able to miniaturize the mechanical processes is one of the most wonderful and mystifying aspects of what Harrison accomplished. Inside, many of the mechanical parts are comprised of diamonds and rubies, painstakingly cut to minimize friction. Another undiscovered secret of H-4 is how Harrison was able to jewel the watch with such minute precision that it exceeded, by several orders of magnitude, any timepiece that had previously been created. Even Harrison said of it: "I think I may make bold to say, that there is neither any other Mechanical or Mathematical thing in the World that is more beautiful or curious in texture than this my watch or Timekeeper for the longitude . . . and I heartily thank Almighty God that I have lived so long, as in some measure to complete it."

One clue has been provided us with regard to the creation of Harrison's No. 4. In the early 1750's one of the artisans whom Harrison had come to know and collaborate with in London, a John Jefferys, made Harrison a pocket watch for his personal use. Jefferys clearly followed Harrison's design specifications, since the watch, which is still extant, contained certain inventions that only Harrison had developed. This may have given Harrison the impetus to change his vision of the sea clock to that of a watch.

Before the trials for H-3 could begin (they had been detained because of the Seven Years War), Harrison presented H-4 to an awed and incredulous Board of Longitude, and sea trials were then proposed for H-4. In a voyage from Portsmith to Port Royal, Jamaica in late 1761 and early 1762, it was determined that H-4 lost only 5 seconds after 81 days at sea. On the voyage back from Jamaica it encountered terrible weather. Rough seas submerged the decks and soaked the watch. But on its arrival home H-4 was still ticking, and its adjusted total error on a voyage lasting nearly six months amounted to just under 2 minutes, well within the margin required for the prize.

It seems clear that the longitude prize should have gone to John Harrison then and there. His watch had done everything that the Longitude Act had requested. But the Board itself was made up primarily of supporters of the lunar distance method who were unable or unwilling to appreciate the triumph of technology that had just occurred. A second test was called for, and this time the watch determined the longitude on numerous occasions to a distance of less than 10 miles, which was more than three times more accurate than the terms of the Longitude Act required.

This success led to political complications. The Board, realizing the significance of what Harrison had done, required that Harrison hand over to them all his sea clocks, plus a full disclosure of the mechanisms inside H-4. In addition, fearing that Harrison's work might never be duplicated without his involvement, the Board of Longitude said that he could not have the prize until he supervised the production of at least two duplicate copies of H-4.

Having done everything the Board had asked, in the fall of 1765 he at last received £10,000 from the Board. This was not the prize. In fact John Harrison was never formally awarded the prize. In 1773, after having completed his new watch, Harrison's No. 5, and having convinced King George III of his entitlement to additional monies, the King saw to it that his Prime Minister, Lord North, engineered the passage of a resolution whereby Harrison received an additional £8,000. This was not the prize either, since it was awarded not by the Board but by Parliament.

John Harrison died on March 24, 1776. Despite the rise in interest in the lunar method for finding longitude in the middle of the century, the success of Harrison's clocks and watches turned the tide. Harrison must have felt vindicated in 1775 when Captain James Cook returned from his second voyage with praise for the method of finding longitude through the use of a chronometer. He was not using Harrison's watch, but one whose technology had been inspired by Harrison's inventions. What remained was to be able to produce watches of comparable precision at reasonable prices. Harrison's watch, as the Board of Longitude continually reminded him, was far too complex for easy, inexpensive reproduction. In fact, the watch that Captain Cook took on his second voyage cost £500. But it was not long before enterprising young watchmakers found a way to replicate the key aspects of Harrison's watch. By 1785 reasonably inexpensive chronometers were being manufactured by several watchmakers in England. And precision chronometers were thereafter available to most sea captains.

Today you can find four of Harrison's time pieces, H-1 through H-4, at Flamsteed House at the Greenwich Observatory. H-5 is maintained at the Guildhall in London. For some historians, Harrison is credited with a large share of ensuring Britain's mastery over the oceans and consequently with the creation of the British Empire, for it was in no small part by means of his chronometer that Britannia ruled the waves.

The Round Table Presentation | November 2, 2000

The Face That Launched 1,000 Quips
Groucho Marx

There are few persons, real or fictional, whose "face" is more universally recognized than that of the person this paper is about. And, in the world of 20th Century entertainment, there is probably no one who possessed a greater gift for the spontaneous, hilarious, often insulting ad lib than Julius Henry Marx.

By the time of his enormously successful second movie - Animal Crackers in 1930 - Groucho Marx had already been in vaudeville and on the legitimate stage for 25 years. The persona that he had created and would reprise in more than a dozen movies as Captain Spalding or Otis B. Driftwood, or Rufus T. Firefly, had been developed from continuous experimentation before thousands of audiences. And, at the time of that second movie, Groucho's career was to extend for yet another 35 years in film, on radio and television.

Julius was born in 1890 on New York's upper east side, the third of five sons of Minnie and Sam Marx. His two older brothers were Leonard and Adolph. His two younger brothers were Milton and Herbert. Julius was considered the brightest of the brood, somewhat withdrawn, and an avid reader of the cheapest entertainment of the era – Pulp Magazines. Refusing to be drawn into the rough life of the streets by his two older brothers, he had early aspirations of becoming a doctor.

The father, Sam, attempted to make a go as a tailor, but possessed neither the skill nor the perseverance, and he failed to prosper. The parents, to make ends meet, encouraged the young boys to work, and Minnie, their mother, began her relentless push for the boys to enter show business. Her own father had been a traveling magician in Germany. She purchased a piano and made the three older boys learn to play and sing.

At 15 Minnie took Julius out of school and made him audition to become the third member of the Leroy Trio. He won the part and left with the Trio on the vaudeville circuit. Opening in Grand Rapids, they played cities all the way to Cripple Creek, Colorado, where Leroy and the baritone, who by then had apparently become lovers, absconded with the act and all of Julius' money. He had to work his way home to New York, where he promptly found another role, this time with an English performer named Lily Seville. They were billed as Lady Seville and Master Marx. Playing through towns across the old south, they ended up in Texas where Lily proved as perfidious as Leroy, leaving unannounced with Professor Renaldo, the animal trainer in another act. This trip Julius had carried a grouch bag with him, so-called because it contained ingeniously hidden compartments where he kept his money. Alas, Lily and the Professor had discovered them all and taken everything he had. Groucho returned home discouraged, but his mother pressed him to try once more.

His third effort was a charm. Auditioning for the Gus Edwards Troupe (Edwards' songs include *By the Light of the Silvery Moon, In My Merry Oldsmobile* and *School Days*), he toured the country as one of the Messenger Boys, itself one of several acts in the Troupe. Others touring as part of the Gus Edwards Troupe included George Jessell, Eddie Cantor, Ray Bolger, Eleanor Powell, and Walter Winchell. The highlight of this tour occurred in May 1907 when Julius shared the stage of the Metropolitan Opera House with Enrico Caruso and Ignace Paderewski raising money for the victims of the San Francisco earthquake.

Julius now became the centerpiece for his mother's plans to bring together all the brothers (except Herbert, the youngest) in a new "kid" act. Kid acts were all the rage in the first decade of the century. There were the four Cohans, the Three Keatons, featuring Buster Keaton, the Rich Brothers, and the Allen Sisters starring Gracie Allen. Minnie put together Leonard, Adolph, Julius and Milton and called them The Nightingales. They were billed as singers. But vaudeville was always a crucible for experimentation. It was low theater and the audiences usually had low expectations of the performers. The boys soon found that cracking a few jokes and ad libbing between their songs brought much laughter and that, in fact, audiences tended to find them extremely funny. Especially Julius. They began buying some material and writing short routines of their own that they inserted in their act. The skits always involved a lot of mischief and misbehaving which the boys would often celebrate by singing stanzas of a bit of doggerel called Peasie Weasie. This was pretty elemental stuff, but it got a lot of laughs. And it gives an early example of the wordplay that became a hallmark of Groucho's comedy:

My mother called sister downstairs the other day.

"I'm taking a bath" my sister did say.

Well, slip on something quick, here comes Mr. Brown.

She slipped on the top step and then came down.

CHORUS

Peasie Weasie, that's his name.

Peasie Weasie, what's his game?

He will catch you if he can,

Peasie Weasie is a bold, bad man!

Went fishing last Sunday and caught a smelt.
Put him in the fire and the fire he felt.
Of all the smelts I ever smelt,
I never smelt a smelt like that smelt smelt.

(ANOTHER CHORUS)

And so on and so on. I suppose when you start out with material of this quality there's lots of room to improve.

From their coalescence as a kid act in 1908-09, and for the next 15 years until they hit Broadway, the boys were almost constantly on the road. They now began to bill themselves as the Four Marx Brothers – Leonard, Adolph, Julius and Milton. Herbert, the baby, joined them later. Here were four teenagers playing 4 to 5 shows each day, 30 shows a week, 45 weeks a year, all over America. Adolph remembers the experience as a montage of railroad waiting rooms, dreary boarding houses, and small town theaters with unresponsive audiences. Groucho's recollections were that he and his brothers had forsaken formal education and had absolutely no marketable skill outside of singing and comedy, and they were prepared, indeed they had no choice but to, ride that skill as far as it could take them.

As a boy, S. J. Perelman, a highly regarded comedy playwright who later wrote several screenplays for Marx Brothers movies, encountered the Marx Brothers in a performance in Providence, Rhode Island, when he was a boy. Writing about the experience 20 years later, he wryly recalled the other acts in the program: "Fink's Trained Mules, Willy West & McGinty in their deathless housebuilding routine, Lt. Gitz-Rice declaiming 'Mandalay' through a pharynx swollen with emotion and coryza, and that liveliest of nightingales, Grace Larue."

The brothers continued to evolve as an act. Reducing the musical numbers, they created more elaborate skits, and their individual characters began to take form. Julius had the quickest tongue, a gatling gun delivery and an expressive face with his eyebrows waggling. A mustache was suggested by a broad swatch of greasepaint above his upper lip. The irreverent comment, the insult, the cynical statement became part of his persona. The oldest brother, Leonard, played a fair piano and gravitated more and more to the character of a bumbling Italian rascal. Adolph, the second son, who had become quite skilled on the harp, found that the audiences responded more to his physical comedy than the words he spoke. Gradually, his lines became fewer and fewer, and in 1913, after an editorial praised his gift of pantomime and noted that his routine was spoiled by the few lines he uttered, Adolph never spoke in a performance again. Milton, the fourth son, was the bland and smiling juvenile. He was later replaced by his younger brother, Herbert, who adopted generally the same character.

The year 1913 proved significant to the careers of the Marx Brothers in a wholly unexpected way. During a break between shows at a theater in Galesburg, Illinois, all four Marx Brothers sat down to a poker game with Art Fisher, a traveling monologist. The subject of nicknames arose. There was a popular comic strip at that time called Sherlocko the Monk, and it became popular to tack an "O" onto people's names. You can see this coming. Apparently Fisher went around the table, looking at Julius and the grouch bag he always carried, and dubbed him Groucho. Leonard, a perpetual ladies man, he called Chicko which became Chico as more in keeping with his Italian character. Adolph, the harpist, became Harpo. Milton, a hypochondriac, wore gumshoes when it rained and was called Gummo. Herbert's fascination with Zeppelins was responsible for his nickname.

Now the brothers had new names, distinctive characters and some very funny material, much of which was ad libbed verbally by Groucho or in pantomime by Harpo. In fact, their shows were so enthusiastically received wherever they went that they became the proverbial tough act to follow. W. C. Fields, when they were booked together in Columbus, Ohio, wrote, "They sang, danced, played the harp and kidded in zany style, were vaudeville entertainers. Never saw so much nepotism or such hilarious laughter in one act in my life. The only act I could never follow." The story goes that Fields told his manager following one of their performances: "[Those Marx Brothers] You see this hand? I can't juggle any more because I've got noxsis on the conoxis, and I have to see a specialist right away." Jack Benny had a similar reaction: "My God, they did 35 minutes of their stuff and when my quiet act followed, it was a disaster! I just can't describe their act, but after a while I used to stand in the wings before I went on and laugh like hell; which meant I stopped worrying about my act."

The chaotic world the Marx Brothers created on stage did not proceed from any specific plan or idea. It resulted from thousands of shows where they tried out new bits to see if they got a laugh. As Groucho said, "I was just kidding around one day and started to walk funny. [you know, low to the ground] The audience liked it so I kept it in. I would try a line and leave it in too if it got a laugh. If it didn't, I'd take it out and put in another. Pretty soon I had a character." Pretty soon probably covers 10 years. And that went for Harpo and Chico, too. Gummo, and his replacement, Zeppo, served as the straight men for the clowning of the three older brothers. In fact, Gummo left the act while it was still on the vaudeville circuit and started his own business in New York. Zeppo stepped in but by the mid-30s he decided that his character had run out of steam, and he left the group to become a Hollywood agent.

With the decline of vaudeville in the early 1920s, the Marx Brothers moved from the thin plot lines of their vaudeville skits to the so-called "legitimate stage." Many of the successful vaudevillians were moving in that direction: Eddie Cantor, Al Jolson, Ed Wynn and George M. Cohan. The success of the Marx Brothers on Broadway was sealed as a result of three plays, written uniquely for them and the characters they had invented: I'll Say She Is, The Cocoanuts, and Animal Crackers. All of these were extraordinary hits and assured their fame. What made the performance of these plays so unusual and exciting was their spontaneity. Every night, the Brothers would abandon the script for long periods and let the repartee and the ad libs and Harpo's physical comedy, take them in any direction it might. And it was Groucho's irreverence and outrageous insults that the audience seemed to enjoy the most. On election eve in 1924 Groucho remarked to Chico, "I see we have the Honorable Jimmy Walker here." He then addressed the Mayor in the audience who was at that time under investigation. "What are you doing here? Why aren't you out stuffing ballot boxes?" The night that President Coolidge attended The Cocoanuts, Groucho interrupted the play to inquire, "Isn't it a little past your bedtime, Cal?" So many deviations from the script occurred that George S. Kaufman, the playwright, was heard to exclaim, with his hand cupped around his ear, "I may be wrong, but I think I just heard one of the original lines."

One of the happy consequences of these extreme departures from the script was that play-goers returned again and again because they knew each performance was going to be different. And the Marx Brothers seemed never to disappoint their audiences.

The fame of the Marx Brothers on Broadway had an unexpected consequence that bears special mention among this group. Harpo became an intimate friend of Alexander Woollcott and was invited to become a member of the famed Round Table at the Algonquin Hotel. The irony was that this child of the streets of New York who, by his own admission, was only semi-literate, would become accepted, even celebrated, by what many agreed was the most sophisticated literary circle in America. What's more, Harpo was also sought out by and became friends with such luminaries as George Bernard Shaw and Somerset Maugham. There was clearly more to Harpo than his on-stage character would ever suggest.

With the advent of talking pictures, the Marx Brothers quickly and easily jumped into this new medium where at least three of the brothers (Groucho, Chico and Harpo) were to make a total of 13 films. While all of these movies contain a mine of rich material, the consensus seems to be that the first seven films are the classics: The Cocoanuts, Animal Crackers, Monkey Business, Horse Feathers, Duck Soup, A Night at the Opera, and A Day at the Races. While the antics of Harpo and Chico were much applauded, it was really Groucho's gift at wordplay, his use of puns, his nonsequiturs, his unbelievable effrontery, all said with that big cigar in his mouth and his eyebrows wildly palpitating, that made the films so popular and so unforgettable.

Many of Groucho's funniest lines were spoken to Margaret DuMont. She was one of the stalwarts of nearly all of the Marx Brothers' movies, and her dignified, dowager-like presence proved the perfect foil for Groucho's wit. In Animal Crackers, Groucho, as Captain Jeffrey Spalding, is speaking to two women, one of whom is Margaret DuMont. He makes a blanket offer of marriage and DuMont protests, "Why that's bigamy!" Groucho agrees: "Yes, and it's big o' me, too." Later in a reminiscence of Africa to DuMont, Groucho speaks one of his more celebrated lines: "One morning I shot an elephant in my pajamas. How he got in my pajamas I don't know." Then he elaborates: "We tried to remove the tusks, but they were embedded so firmly we couldn't budge them. Of course, in Alabama, the tuscaloosa, but that's entirely irrelephant to what I'm talking about." And in A Night at the Opera Groucho, as Otis B. Driftwood, hurls one of his insults at DuMont and high society when he remarks, "The strains of Verdi will come back to you tonight, and Mrs. Claypool's check will come back to you in the morning."

Groucho's comedic gifts were extolled by S. J. Perelman, who, as mentioned earlier, had become a very successful playwright and who had a gift for parody, a high intelligence and a rich vocabulary. Perelman and Groucho worked together on several scripts, and he commented:

> *I loved [Groucho's] lightning transitions of thought, his ability to detect pretentiousness and bombast, and his genius for disemboweling the spurious and hackneyed phrases that litter one's conversation. And I knew that he liked my work for the printed page, my preoccupation with cliches, baroque language and the elegant variation. Monkey Business demonstrated his love for word play, including the line 'Love flies out the door when money comes innuendo;" or "I knew a man who had more women than you could shake a stick at, if that's your idea of a good time," or the come-back when the captain of the ship threatens to throw him in irons, "You can't do it with irons. It's a mashie shot. It's a mashie shot when the wind's against you, and if the wind isn't, I am."*

What the Marx Brothers brought to these early movies was disorder, their particular brand of disorder. Chico and Harpo were constantly trying to crash parties, stow away on boats, assume false identities, escape from the arms of the law, randomly chase blondes down hotel or steamship corridors. One writer commented that compared to the world the Marx Brothers created in their films, Alice in Wonderland is a serious book of reference. The disorder Groucho created was through the spoken word. And you could see in Groucho an effort to subvert every attempt at etiquette or rational conversation. As one writer said, "Groucho could never let any stable situation exist. It had to be disrupted as soon as it began to solidify." An example is the following exchange with Margaret DuMont as Mrs. Teasdale:

GROUCHO: I suppose you'll think me a sentimental piece of fluff, but would you mind giving me a lock of your hair?

MRS. TEASDALE: (Coy) A lock of my hair? Why I had no idea.

GROUCHO: I'm letting you off easy. I was going to ask for the whole wig.

In *Horsefeathers* where Groucho plays the new President of Huxley College, Professor Quincy Wagstaff, he makes a similar assault on the conventions of academia. Explaining to the faculty that it's not possible to support both a stadium and a college, he tells them, "Tomorrow we start tearing down the dormitories." Stunned, the faculty members exclaim, "But, professor, where will the students sleep?" Groucho responds, "Where they always slept. In the classroom."

Groucho's celebrated wit was not limited to his on-screen performances. Devotees have collected numerous anecdotes of his famous off-screen quotes. Such as the time he wrote an author, "From the moment I picked your book up until I put it down I was convulsed with laughter. Some day I intend reading it." Or when he applied for a membership at a private club in Long Island, but was told that the club would not allow Jews to swim in its pool. He wrote the club: "My daughter is only half Jewish. What if she only goes in up to her waist?" Or when he wrote to a friend: "The secret of life is honesty and fair dealing. If you can fake that you've got it made."

By the end of World War II, Groucho was 55 years old. The time of the Marx Brothers films and their raucous, insolent brand of comedy had run its course. But while his brothers were settling into retirement, Groucho deftly jumped to a new medium – radio. In 1947 he hosted the radio program

"You Bet Your Life." What made the show so popular was the unpredictability of Groucho's ad libs. Groucho's earlier experiments with radio programs had failed, some believed, because he was confined by a script. But on "You Bet Your Life" Groucho was freed from a script and was allowed to improvise. The result was that by 1950 "You Bet Your Life" had risen to sixth place among all network radio programs. Groucho had found a new vehicle and post-war America had discovered Groucho.

In 1950 Groucho signed with NBC to take "You Bet Your Life" to television where the show ran for 11 years. For the TV version of "You Bet Your Life," network executives wanted a return to the greasepaint mustache and claw-hammer coat. Groucho adamantly refused, but did consent to grow a real mustache. I'm sure each of us has favorite recollections from the TV show, with Groucho sitting on a stool, chatting up guests, joking with George Fenneman, and the duck that looked like Groucho dropping into the picture when someone said the secret word. One exchange on the program that didn't survive the editing process, but which has become legendary nonetheless, was Groucho's comeback to a woman who proudly declared that she had ten children because she loved her husband. Groucho replied, "I love my cigar but I take it out of my mouth once in a while."

Groucho's last season of "You Bet Your Life" ended in 1961. Thereafter he did numerous stints on "The Tonight Show," and took to writing, publishing several books during the 1960s.

In 1972 a celebration of Groucho's life and humor was held at Carnegie Hall with Groucho, age 82, as the honored guest, and there was much to celebrate. Groucho had appeared on the cover of Time twice. He had authored or co-authored nine books and one quite successful play. The Library of Congress had requested his private correspondence to be placed in its national treasures collection. His remarks had appeared in Mencken's *The American Language*, as well as *Bartlett's Familiar Quotations*, and *The Oxford Dictionary of Quotations*. It was Groucho that Winston Churchill was watching during the heaviest air raid of the Battle of Britain. T. S. Eliot, at the time the world's pre-eminent poet, had written Groucho requesting an autographed picture and followed with an effusive thank you letter when Groucho obliged him. At present there are no fewer than 20 websites offering pictures, quotes, stories and merchandise relating to Groucho. He is the subject of nearly 30 books. Comedians such as Woody Allen and Jerry Seinfeld hold Groucho up as the gold standard, the father of modern comedy, praising his sense of rhythm, and noting that you can hear his influence in the delivery of every standup line spoken today.

Groucho Marx died on August 19, 1977. Like Jack Benny and George Burns, Groucho was a rarity, one of those few comedians whose genius had allowed his career to span the history of entertainment in the 20th Century: vaudeville, theatre, film, radio and television. His face and his humor will influence, and are destined to be enjoyed by, succeeding generations, and Groucho would hopefully appreciate the irony of this parting comment – that at least in one respect the 20th Century has witnessed the triumph of Marxism.

The Round Table Presentation | December 5, 2002

It's Only A Matter Of Time...
Exploring The Characteristics And Mysteries Of Time

This paper is about time, and I suppose the notion of writing it originated from several sources. Four years ago I presented a paper on longitude at sea and John Harrison's clocks. Then a little more than a year ago, Brad Gioia presented a paper on Andrew Marvell's poem "To His Coy Mistress" at Bob McNeilly's home. The allusions to time were central to the poem: "Had we but world enough and time . . . " and "time's winged chariot." Those items, together with the fortunate confluence of a recent Scientific American issue that contained some fascinating articles dealing with the concept of time, led me to the research and writing of this paper. And of course, I can't deny the fact that as a lawyer I have had a professional interest in timekeeping, timesaving, and timeliness for, I don't know, time out of mind.

So I thought it might be interesting to explore various aspects of the subject of time. In science, psychology and everyday life.

Let me begin by establishing a frame of reference, a look at the units of time by which we measure our world. Here is a sampling of various units of time ranging from the briefest of moments to the longest span of cosmological time.

1 attosecond (a billionth of a billionth of a second). The briefest events that scientists can measure are expressed in attoseconds. Using high speed lasers, pulses of light have been created that last just 250 attoseconds. This time interval, while unimaginably short, is nothing compared with what is called Planck time (named for the physicist, Max Planck). Planck time is the time it would take a photon to travel a distance known as a Planck length, which is equal to 10-20 times the size of a proton. We are talking "brief." In case you were wondering, Planck time is the briefest period for which Einstein's theory of general relativity is valid as a classical theory of gravity.

1 femtosecond (a millionth of a billionth of a second). An atom in a molecule typically completes a single vibration in 10 to 100 femtoseconds. The interaction of light with molecules in the retina - the process that allows vision - takes about 200 femtoseconds.

1 picosecond (a thousandth of a billionth of a second). The fastest transistors operate in picoseconds.

1 nanosecond (a billionth of a second). The microprocessors inside the newest personal computers will typically take 2 to 4 nanoseconds to execute a single instruction, such as adding two numbers. It takes 3 nanoseconds for an atomic bomb reaching critical mass to detonate.

1 microsecond (a millionth of a second). A beam of light will travel 300 meters in 1 microsecond. The flash of a high-speed commercial stroboscope lasts about 1 microsecond. Because of the relativistic effects of gravity, clocks atop Mt. Everest outgain clocks at sea level by about 30 microseconds a year.

1 millisecond (a thousandth of a second). The shortest

exposure time in a typical camera. A housefly flaps its wings once every 3 milliseconds. In computer lingo, an interval of 10 milliseconds is known as a "jiffy."

1/10th of a second – the duration of the fabled "blink of an eye." A hummingbird can beat its wings seven times in 1/10th of a second.

1 second – a healthy person's heartbeat lasts about this long. On average, Americans eat 350 slices of pizza each second. Historically, a second was the 60th part of the 60th part of one hour. Science is more precise: it is the duration of 9,192,631,770 cycles of radiation from a cesium 133 atom.

1 minute. The average person can speak about 150 words in a minute. Light from the sun reaches us in about eight minutes.

1 hour. Reproducing cells generally take 1 hour to divide. One hour and 16 minutes is the time between eruptions of Old Faithful at Yellowstone Park.

1 day. This is our most natural unit of time. The human heart beats approximately 100,000 times in a day.

1 year. It takes 4.3 years for light from Proxima Centauri, the closest star, to reach Earth.

1 century. Baby boomers have only a one in 26 chance of living to the age of 100, although members of the Round Table have considerably better odds than that. The great oaks of the British Isles have a life span of about four centuries.

1 million years. A spaceship moving at the speed of light would not yet be at the halfway point on a journey to the Andromeda galaxy, which is 2.3 million light years away.

1 billion years. This is the time it took for the newly formed Earth to cool, develop oceans and witness the creation of single cell life. The Universe is between 12 billion and 14 billion years old. Some cosmologists believe that the Universe will keep on expanding indefinitely, until long after the last star dies which is expected to be 100 trillion years from now. Our future stretches ahead so much farther than our past trails behind.

Source: *Scientific American*, September 2002 issue, pages 56-57

These various units of measured time bear witness to the fact that time has become the most accurately measured of all phenomena. Keeping and marking time is a practice that undoubtedly goes back into prehistory, and periods of time expressed by the solar and lunar cycles came to be measured with some accuracy by the Egyptians and Babylonians and the Druids. At the time of Rome's ascension to power in the Mediterranean, sundials and waterclocks were widely used, but even then, timekeeping had its detractors. As Titus Plautus wrote in about 200 B.C., "The gods confound the man who first found out how to distinguish hours. Confound him, too, who in this place set up a sundial, to cut and hack my day so wretchedly into small portions."

Precision timekeeping really began with the development of the mechanical clock in the 13th Century and continued as a mechanical process until earlier this century when battery power and quartz crystals combined to create an extremely accurate, low power clock or watch. But Quartz crystals are the minor leagues when we look at the so-called ultimate, or atomic, clocks. These are clocks that are not mechanical (like a spring) or electromechanical (like a quartz crystal).

They are quantum mechanical. Electrons of cesium atoms will change their magnetic field when hit with a blast of microwave radiation at a frequency of 9,192,631,770 cycles per second. The frequency of oscillations of cesium atoms is stable and unvarying. That exact number of cycles in a cesium atom is the standard measure of the second. With cesium clocks the second can now be measured to an accuracy of 14 decimal places, a precision that far outstrips any other fundamental unit of measure.

It is clear that so many of our technological advances have been made possible because of our ability to measure time accurately. Television transmitters must have atomic clocks, for if the television frequency is not absolutely precise, the picture will flip up and down, as we used to see 30 years ago before atomic clocks eliminated that pesky annoyance. Cellular phone networks need atomic clocks to squeeze more and more channels of communication into precisely tuned bandwidths. In addition, our Global Positioning System depends on precision timekeeping. The GPS system is comprised of a series of satellites in geosynchronous orbit that constantly broadcast their exact whereabouts and the exact time measured by synchronized atomic clocks maintained on board each satellite. Receiving devices process this information from no fewer than four satellites at a time into precise terrestrial coordinates whether it's for a hiker in the Cascades, an aircraft flying over the Atlantic, or the driver of a top-of-the-line Mercedes. An error of one millionth of a second in any of those atomic clocks could mean a discrepancy of more than one-fifth of a mile. And of course our military relies on precise timing for command and control of cruise missiles as well as laser delivery of airborne explosives. It is no coincidence that the Directorate of Time (that has an Orwellian sound to it) at the U.S. Naval Observatory is part of the Department of Defense. You can obtain U.S. Naval Observatory master clock time on the Internet by searching under the Directorate of Time.

This scientific obsession with accurate timekeeping is fully reflected in our everyday lives, certainly in the United States. If we define time as a continuum in which events occur, I think we would all agree that the number of events that each of us squeezes into a given time interval has increased at a breathtaking pace. We witness people who have dedicated their lives to cramming as many experiences as possible into a set amount of time as if somehow they will come out ahead as a result.

I recall when my law firm in Atlanta first obtained word processing equipment in the late 1960s. At last proofreading could be minimized. We would all have more time to give more thought to the legal issues in those contracts. This would benefit the clients and they would be pleased. It didn't work out that way. Once our clients realized that we could turn around documents in one day rather than three, they began to expect one-day responses. Now, with email, the response time is more like one hour. What a foolish notion it was to think that these time-saving devices would give us more time to consider.

And of course that experience was only the beginning. Remember the Federal Express advertisement that suggested that what we really needed was the opportunity to have a package delivered to us "absolutely, positively overnight"? Delivery in two or three days didn't seem to bother us 30 years ago.

Other examples abound. Our children's days are intensely programmed. Ours rarely were. We jab the "door close" button several times in frustration to hurry the elevator on its way. We need speed dial buttons on our telephones. Multi-tasking has become the touchstone of our obsessive drive to become busier and busier. And then, as if this pace were not fast enough, we now have the Internet. We no longer have the need to wait for the Fed Ex package. Documents can be transmitted in minutes. We used to deal with a time scale of days or hours; now we operate and expect information on a time scale of minutes. And with the Internet and news networks like CNN there is also an immediacy to the experience of events that simply didn't exist in the middle of the 20th century. Then, information about events from the far side of the world would be revealed to us after a respectable delay, a time delay that made sense to all of us because, after all, it happened 10,000 miles away. Now we must become accustomed to experiencing the simultaneity of events both near and far.

Sophocles remarked that "time is a gentle deity." Maybe it was for him, but these days it cracks the whip![1]

One of the central mysteries regarding the nature of time really has a dual aspect to it: Why does time seem to flow and why does it flow only in one direction? Time's arrow points only one way, forward. For us the past is fixed, unredeemable, the present "now" is our reality which, even as I speak, has just slipped into the past, and the future is unknown to us. Time moves relentlessly on a one-way street from the past into the future. Why is that so?

Most modern physicists reject the notion that time is constantly flowing, but prefer to think of time as laid out in its entirety – a sort of timescape analogous to a landscape. In such a timescape all past and future events are located at some place on the continuum of time and space. Under this construct, time does not flow, it just is. It would be nonsense to talk about the speed at which time passes. According to this theory it doesn't pass at all. What we witness as time's flow is not related to time itself but rather to the different states we find the world in. Time is the context within which we order these different states. The states change but time simply exists. If this hypothesis about a timescape

[1] Faster: The Acceleration of Just About Everything, by James Gleick, page 13.

is correct, then movement back and forth in time is a logical possibility. And yet while we can imagine a notion like this and we accept its logic, our experience tells us it can't be so. We experience the flow of time and we experience it in one direction only.

The prevailing explanation why we perceive time's arrow to be pointed only in a forward direction is based on the second law of thermodynamics, which we know as entropy. The law of entropy states that a closed system evolves to a state of maximum disorder over time. We are all familiar with TV ads showing an egg shattering on the kitchen floor. The hook associated with those ads is that the camera reverses itself so that the egg reforms and flies up into the man's hand. We have never witnessed this happen in reality, and we never expect it to happen. A broken egg has a higher entropy than its unbroken counterpart. And for us this process cannot be reversed.

Stephen Hawking addressed this question in A *Brief History of Time*. According to Hawking we perceive time to flow in one direction because it is in the same direction as information storage in our brains. Our memories accumulate knowledge in only one way - from the past to the future. Psychologically we can only comprehend time as moving in that direction. Hawking links this memory storage to entropy arguing that our bodies, including our brains, are subject to the second law of thermodynamics. In other words, entropy is increasing in the same direction as our life processes, which would include accumulation of memory. For that reason we are incapable of perceiving a reversal of time's arrow, even if we can imagine it.

It is interesting that the equations of Newton and Einstein as well as the equations of quantum mechanics, work independently of time's arrow. That is to say they are equally valid whether time is presumed to move forward or backward. Einstein confirmed his disagreement with the world's perception of time when he said, "The past, present and future are only illusions, even if stubborn ones." And in quantum mechanics, physicists have long subscribed to the notion that at the level of interaction of subatomic particles, time may in reality move both forward and backward. But while this may be true for subatomic particle interactions, when these particles are assembled into the materials that comprise the physical world that we know and understand, time's flow appears only to move in one direction.

While time's forward arrow continues to mystify, it had long been a universally held notion that time is nonetheless absolute and the same for everyone and everything everywhere. This long-held belief was completely refuted by Albert Einstein. Einstein proved that an event could occur at one moment in one frame of reference and at a different moment in another frame of reference. In other words, time was not the same for those two frames of reference. It was relative. This assault on the absolute nature of time was expanded under Einstein's general theory of relativity where, as a result of acceleration, he showed that time slows for an accelerating body as compared to a body in uniform motion. Atomic clocks placed on spaceships and on aircraft have recorded that this time dilation in fact happens. Moreover, under the general theory of relativity, Einstein established that the force of gravity, which operates to create acceleration in bodies affected by it, itself slows time. A clock in the attic will move slightly faster than a clock in the basement. This also has been well established by experiment.

Time travel has always been a favorite of science fiction writers. H. G. Wells wrote The Time Machine in 1895, nine years before Einstein advanced his theory of relativity. While Einstein's theory of relativity does not permit time travel into the past, it does allow for time travel into the future. One of the most well known of his so-called "thought" experiments is that of the paradox of the twins. Imagine identical twins, one of whom remains on earth while the other travels in a spaceship from Earth to a nearby star and returns. Assume that the space traveler twin accelerates for both the outward and return journey to about 90% of the speed of light. For the traveling twin, the elapsed time of the trip may have been one year, but he finds when he returns to earth that 10 years have elapsed and that his brother now is nine years older than he. The travelling twin has voyaged nine years into the future.

Since a person travelling at the speed of light would find that time essentially stood still as compared to another person who was stationary, some scientists theorize that a person travelling at speeds in excess of the speed of light would actually travel back in time. Of course, Einstein's theory posits that nothing can travel faster than light, but scientists today theorize that wormholes in space, created in black holes, could be a means of travel back in time. There are plenty of interesting books and articles on this notion, and if you saw the movie Contact, based on the Carl Sagan novel, this time travel was depicted. It has also inspired some humorous reflections, as in the following limerick:

There was a young woman named Bright

Who could travel much faster than light

She went out one day

In an Einsteinian way

And returned the previous night

Now let's move from physics to biology. The biological timepieces that exist in nature abound: cicadas swarming every 17 years; the swallows returning to Capistrano or the buzzards to Hinkley, Ohio on the same dates every year; or the migration of the Monarch butterflies. All organisms have biological clocks. Some are precise and predictable; others are less regular. Let's briefly look at three such timing mechanisms in the human body.

The biological timepiece of which we are most aware is our circadian clock. Attuned to the cycles of light and dark, this clock programs our sleeping habits (releasing melatonin when darkness occurs). It adjusts our blood pressure, body temperature and regulates other bodily functions. Investigators initially targeted a point in the brain as the physiological mechanism that operated our circadian clock. Experiments in the mid 1990s, however, discovered four genes that also are intimately involved in regulating our circadian rhythms. It appears that these genes are expressed in every cell of the body. Studies show that these circadian patterns continue even in the absence of a change of daylight and darkness. In a study of human cells in a petri dish under constant lighting, the cells continued to follow regular 24 hour cycles of activity, including hormone secretion and energy production.

The human body contains another timing mechanism that is quite different from the slow but regular pattern of the body's circadian clock in that it regulates split-second movements. Named the interval timer and linked to a portion of the brain called the striatum, this precision timepiece tells us just when to hit that tennis serve or how fast we much run to catch a pass. It allows us to clap in time to the beat of a song that we like, or to decide whether or not we can make it through a yellow traffic light before it turns red.

Unlike the circadian clock, this interval timer enlists the higher brain functions as we make judgments about the time it takes to perform these split-second actions. And it improves with practice as our memories store data of these repetitive actions. Musicians and athletes are proof of this.

And finally, we have the body's so-called clock of life. As we eliminate disease the question becomes whether immortality or perhaps a lifetime rivaling Methuselah, is possible. Scientists who have studied the aging process have looked at many factors that could lead to longer lives. Where there is some general agreement about a factor that may determine life's length comes from studies at the cellular level. I am referring to mitosis or cell division. There is considerable evidence that cells have a given number of divisions in them and then, like sand passing through an hourglass, there are no more. These cells don't die, but they do stop dividing. They go into permanent retirement, become less resistant to viruses or bacteria, and when this happens, let's say in the immune system, well, as my paper is entitled, it's only a matter of time.

Now let's move from our body's clockwork to some observations about how the mind experiences time. One of the most fundamental and persistent questions regarding psychological time relates to why is it that time seems to flow at different rates for us depending on whether we are engaged in a pleasant activity or whether we are bored? One of the wonderful stories, probably apocryphal, relates to a question raised by Einstein along the following lines. Why is it that when a man sits with a pretty girl for an hour it seems like a minute? But let him sit on a hot stove for a minute and it's longer than any hour. That, said Einstein, is relativity.

And that leads to another question: Why is it that time seems to move so much more quickly as we get older? All of us can remember times in childhood when a day seemed endless, where we rarely thought about tomorrow, and where any reference to next year seemed like forever. Why is our perception of time so different now that we are adults? Why do the years now speed by so quickly that we can't differentiate them any longer?

One reason given by psychologists is that for a child of, say seven, one year represents 15% of that child's life. For a

50-year old, on the other hand, one year is only 2% of that person's lifetime. Perhaps the acceleration of time can also be explained by how busy we have allowed ourselves to become as adults. Remember the common complaint from children "there's nothing to do?" Well, we don't seem to have that problem in our adult years. And, finally we must admit, that it may be our sense of mortality, becoming more acute with age, that causes our sense of time to speed up.

So, as I conclude tonight this brief excursion through the various aspects of time in our world, I'm afraid I can't answer the crucial question: What is time? How can it be defined? St. Augustine in the 5th century stated that he knew well what time was until he was asked to describe it. Then he could not find the words. Despite all the advances in knowledge over the past 1,600 years it would appear that scientists are not much closer to an understanding of what time really is. Referring to it as a 4th dimension timescape is helpful but not fully enlightening.

You may have heard the humorous definition of time as being nature's way of insuring that things don't happen all at once. I've used that definition in jest on occasion, but having spent some "time" considering this subject, I think that might serve as an accurate, if not profound, characterization of how we view time. Time serves as a context or a construct that helps us to order and understand our world.

Because we cannot describe motion without a reference to time, perhaps the reverse is true as well – that is, we cannot describe time without motion. If we could imagine a static universe – where there was no motion – then that would be a universe where time itself has no meaning.

The mystery of time continues to be a challenge to scientists, philosophers and theologians. Philosophers have placed great significance on the present moment because, existentially, the present would appear to be the only reality. In the Old

Testament, in Exodus, Chapter 3, God revealed himself to Moses as "I AM THAT I AM." It is striking to me that the God of Israel chose to reveal himself as the great I AM, the eternal present, transcending past and future, and suggesting that the mind of God encompasses the whole of time, but that God operates independent of time. It's an extraordinary image and a fitting way, I think to end this paper. As the writer of Ecclesiastes declared, "To every thing there is a season, and a time to every purpose under heaven."

BIBLIOGRAPHY

Scientific American, Special Issue, September 2002

James Gleick, *Faster: The Acceleration of Just about Everything* (Pantheon Books, 1999)

Paul Halpern, *Time Journeys* (McGraw-Hill, Inc. 1990)

S. W. Hawking, *A Brief History of Time* (Bantam Books, 1988)

Michael Shallis, *On Time* (Schocken Books, 1982)

FUZZY LOGIC

AN EXTREMELY BRIEF HISTORY OF TIME

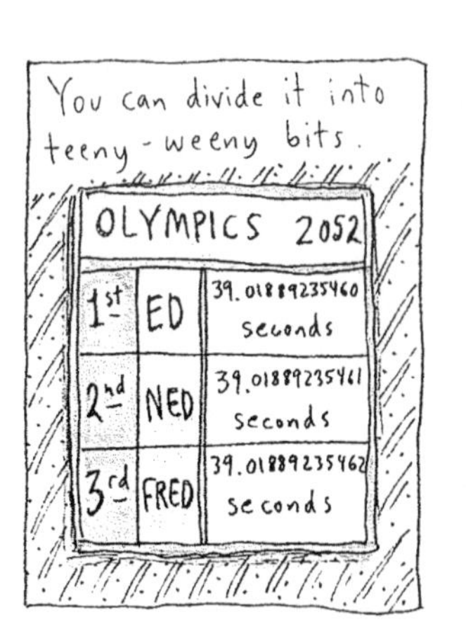

R. Ch

The Round Table Presentation | October 6, 2005

That Certain Feeling
Intuitive Intelligence

Several years ago Vic Braden, a well-known tennis coach and commentator, began to notice something strange whenever he watched a tennis match. What Braden realized was that as a player was initiating his second serve, he always knew when that player was about to doublefault. It didn't matter whether it was a man or woman playing, whether he was watching the match live or on television, or how well he knew the person who was serving. So at a professional tournament in southern California several years ago, Braden began to keep track of his predictions. On the first day he correctly predicted 16 out of 17 doublefaults; the next day 20 out of 20, and this extraordinary ability has continued ever since. Now Vic Braden is a renowned tennis coach so it's not surprising that he should be really good at the subtleties of tennis, and in particular serving. He was seeing something in the way a tennis player would stand, how the toss would occur, or how the player's motion was initiated. Any number of things could allow him to know a doublefault was going to occur. But here's the catch - Braden to this day cannot figure out how he knows.

In 1983 the J. Paul Getty Museum was offered an opportunity to purchase a statue that is known as a Kouros, a sculpture of a nude male youth standing with his left leg slightly forward and his arms at his sides. Kouri were sculpted in Greece in the Sixth Century BC. They are stylized statues, and approximately 200 have been found over the years, most badly damaged or in fragments. This one was not. It was almost perfectly preserved. The proposal came from a wellknown art dealer named Gianfranco Becchina, and the asking price was $10 million. The Getty was a new museum with a great deal of money to spend, but its curators moved very cautiously. Over the next 14 months the curators checked out the art dealer, received documents that established the statue's provenance (or history of the statue's discovery and ownership), and brought in a noted geologist to examine the statue carefully. A core sample was removed from behind the statue's knee, and the geologist confirmed that the statue was indeed made of dolomite marble from an ancient quarry on the island of Thasos, Greece. Moreover, the x-ray diffraction and electron microscope studies confirmed its age and showed that the statue was covered with a thin layer of calcite. This was significant because dolomite marble can turn into calcite only over the course of hundreds if not thousands of years.

The written records showed that the statue had been in a private collection of a Swiss physician since the 1930s when it had been acquired from a wellknown Greek art dealer. In other words, the statue was old and it was authentic. The Getty was satisfied and it made the purchase.

Excited about its new acquisition, the Getty invited several wellknown curators and art experts to view its new purchase. One, an Italian art historian named Federico Zeri, was taken to see the statue, and, as the cloth was removed, Zeri found himself staring at the sculpture's fingernails. In a way he could not articulate, they seemed wrong to him. Next came Evelyn Harrison, one of the world's foremost experts on Greek sculpture. As soon as she saw the Kouros, she asked if they had paid for it. When the curators answered "yes," her immediate comment was, "I'm sorry to hear that." Shortly thereafter, Thomas Hoving, the director of New York's Metropolitan Museum of Art, came to see the statue. Hoving had the habit of making a mental note of the first word that passed through his head when he saw a new piece of art, and he recalled that the first word that went through his mind when he saw the Kouros was "fresh." And he realized fresh was not the right reaction to have to a 2500 year old statue. All three of these individuals had immediate flashes of perception concerning what they saw, though none could articulate what it was that troubled them about the statue. Although the Getty had already purchased the Kouros, based upon their intuitive reactions it began a more indepth investigation into the statue's background. Investigators found numerous discrepancies and falsehoods in the documents reflecting ownership in Switzerland and the earlier transfer from Greece, and this information convinced them that the documents were fakes. But what about the calcite film on top of the statue? A scientist informed the curators that the calcite could be created in a couple of months using potato mold. Now, in the Getty catalog there is a picture of the Kouros with the notation "Kouros: About 530 BC, or modern forgery."

Zeri, Harrison and Hoving each felt an "intuitive repulsion" about what they were seeing in their first glance at it. In that first moment they were able to understand more about the essence of the statue than the team at the Getty was able to understand after 14 months of investigation, though none of them could explain why they felt as they did.

As Jackie Larsen left her Grand Marais, Minnesota church prayer group one weekday morning in April 2001, she encountered Christopher Bono as she was walking to her village shop several blocks away. Bono was a short, cleancut and wellmannered young man. He told Jackie that his car had broken down and he was looking for a ride to meet friends in Thunder Bay. "I told him to come with me to my shop and I would look up his friends in the phone book and they would come for him," she later recalled.

But as they walked along, Larsen felt a physical pain in her stomach. Initially she thought the young man might be a runaway, but some sense told her that this situation was very wrong. She did not invite the young man into her shop but insisted that they talk outside on the sidewalk. "I said to him 'I am a mother and I have to talk to you like a mother . . . . I can tell by your manners that you have a nice mother.'" Bono responded saying "I don't know where my mother is."

Terminating the conversation, Larsen encouraged Bono to go back to the church and talk to the pastor. When he left her she went inside her shop, called the police and suggested that they trace the license plates on his car which she had seen outside. It turned out the car was registered to Bono's mother in southern Illinois. When police went to the mother's home, no one answered. Breaking in they found blood everywhere and Lucia Bono dead and lying in the bathtub. Christopher Bono, 16 years old, was charged with first degree murder.

Jackie Larsen had a gut feeling. She intuitively sensed that something was wrong. Just as Vic Braden and the art experts had immediate feelings about what they were observing. This paper is about the experiences reflected in these three stories, about the power of intuition that we all have, how it benefits us, and where it leads us astray.

Let's begin with the definition of intelligence which Webster's Dictionary defines as "the capacity for learning, reasoning, and understanding . . . aptitude in grasping truths, relationships, facts, and meanings." Experts today tend to divide intelligence into three different branches: the power of recall or memory; the ability to reason analytically; and intuition. Webster's defines intuition as our capacity for direct knowledge or for immediate insight without detailed observation or reasoning. Intuitive thinking is experiential, simple, rapid, effortless and irrational. By irrational I mean it cannot be explained in a rational, or reasoned, manner. By contrast, rational or critical thinking is logical, analytic, linear, deliberate and slow.

As we intuitively know, we all have intuition, though as our experience teaches us, we have it in greater and lesser measures.

What do we know about intuition? Well, we know that it is rooted in the ancestral instincts of our forebears. Before the invention of language or tools or the organization of social groups, any of which would have required some degree of analytical reasoning, our ancient ancestors survived on intuition. Those who failed to identify threats or who were unable to react immediately to attacks from predators were removed from the gene pool immediately and permanently. Some of our so-called instinctive reactions, expressed often in the form of phobias, were probably hardwired in our DNA over thousands of generations and control a large portion of our behavior today. We all certainly seem to have an intuitive tendency to bond with things that are familiar and to avoid things that are unfamiliar. Today instinct and intuition are often used interchangeably, but I think it is more correct to say that instinct is only one element of our broader, intuitive intelligence.

The truth is that until quite recently the notion of intuition received very little respect. Despite the fact that intuition was the first branch of intelligence used by homo sapiens, over the last several centuries, beginning with the Enlightenment, intuitive intelligence has taken a back seat to critical analytical reasoning. Often referred to disdainfully as women's intuition, it also became identified with the paranormal - mind readers, mystics, and the like. But recent studies over the past 50 years have moved intuitive intelligence back to a place of prominence, if not yet on equal footing with analytic reasoning.

These studies of intuitive "knowing" have revealed a fascinating unconscious region where our minds operate, you might say, backstage and out of sight. This unconscious "knowing" is not in any way related to Freud's "subconscious," which was a dark, nether world of repressed thoughts and fears. This unconscious region operates independently from our conscious mind, has greater flexibility, and yet is able to inform our conscious mind with intuitive judgments and insights. Our mental processes thus operate on two levels – conscious and deliberate on the one hand and unconscious and automatic on the other. In effect we are engaged in dual processing, and as a result we know much more than we think we know.

Contemporary psychological science has concluded that most of our everyday thinking, feeling and acting operate outside conscious awareness. Think about the simple act of speaking. Strings of words spill out of your mouth, almost effortlessly. Much of what is going on is not part of your consciousness. It is as though somebody up there in your brain is organizing the sentences and paragraphs for you, choosing the proper words, putting them into proper grammar and syntax, being sure that they are properly pronounced and then ordering your mouth to speak them. Certainly your conscious mind may have a general notion of what you want to say, but the saying of those words is pretty much automatic. Everyday life reflects hundreds of activities each of us undertakes without thinking about them, whether it's tying a tie, buttoning a shirt, or signing your name. These are automatic actions that seem to be taken without any conscious thought. As one scientist said, "Be glad for the automaticity of being. Your capacity for flying through life mostly on autopilot enables you to function effectively." And as the philosopher Alfred North Whitehead observed in 1911, "Civilization advances by extending the number of operations which we can perform without thinking about them."

Moreover, the unconscious mind seems to have a great

deal more flexibility than our conscious mind. Consider our capacity for divided attention. Our conscious attention is highly selective. It can only be in one place at a time. Have you ever been able to listen to and make sense of two separate conversations, say at a cocktail party? The fact is you can't, though you may hear a word or two from one conversation while your attention is focused on another. And here's a little experiment to try that demonstrates how your conscious mind can only attend to one thing at a time. If you are righthanded, try moving your right foot in a smooth counterclockwise circle while writing the number 3 repeatedly with your right hand. Or in the simpler times in which we all grew up, you could try to pat your head and rub your tummy. Now you can do either one of these things quite easily, but you can't do them at the same time, at least not without some practice. As one author said, "If time is nature's way of keeping everything from happening at the same time, then consciousness is nature's way of keeping us from thinking about everything at the same time."

The unconscious mind, however, works quite differently. Experiments conducted over the last 30 years indicate that our unconscious minds can receive, process and be influenced by enormous amounts of information while our conscious minds are focused on something else. While psychologists and scientists agree that all of that unconsciously received information is stored somewhere in our capacious memory banks, there is general agreement that it is not as easily retrievable as things to which we have addressed our conscious mind. Nonetheless, all of that unconsciously received information influences our thinking and our actions. It explains why sometimes we intuitively feel something that we do not actually or consciously know. Our minds are processing and digesting vast amounts of information totally outside our consciousness.

Scientists have summarized the two ways of knowing I've just described as experiential on the one hand and rational on the other. Experiential knowing is intuitive, automatic, nonverbal, concise and rapid. Rational knowing is analytic, verbal, slow, and justified with logic and evidence. Rationally we may know that flying is safer than driving. The evidence is clear. But experientially, emotionally, we may feel just the opposite. Similarly, our fear of heights, crowds, or enclosed spaces, indeed all our phobias, are driven by our experiential, intuitive right brain which quite often trumps our logical, analytical thinking left brain.

One of the important ways our unconscious mind informs our conscious thoughts and actions has been described by science in recent years as our emotional or social intelligence. Time Magazine devoted a lead article to emotional intelligence several years ago. This can best be defined as the know-how that enables us to comprehend social situations and manage ourselves in them. All of us know people who have knocked the socks off the SAT and yet are significantly lacking in social sensitivity or judgment, and we all recognize that academic aptitude doesn't guarantee social intelligence. Socially intelligent people are intuitively aware of themselves and others. They are empathetic. They comprehend and respond to the reactions of others in varying social situations. They also tend to cope with the burdens of life without letting their emotions get hijacked by anger, depression or fear. Persons with high emotional or social intelligence can read the reactions of others and respond in skillful ways. Research in the area of social intelligence has resulted in a series of tests and the creation of a scale that measures emotional intelligence. Those scoring high on the emotional intelligence scale are thought to be generally more successful in business or in their professions than those who score lower, even if those who rank lower on the scale possess stronger, left-brained analytical skills. In fact, the absence of any emotional intelligence makes it nearly impossible for a person to function effectively in the world. A neuroscientist from the University of Iowa who has studied thousands of braindamaged patients tells the story of Elliot, a man with strong intelligence and memory. Since the removal of a brain tumor, Elliot has lived without emotion. The scientist said that, "I never saw a tinge of emotion in my many hours of conversation with him. No sadness, no impatience, no frustration." When Elliot was shown disturbing pictures of injured people or natural disasters, Elliot showed, and realized that he felt, no emotion. Like Star Trek's Mr. Spock, he knows but he cannot feel. And since he could not respond to emotional signals, his emotional intelligence all but disappeared. Unable to adjust his behavior to respond appropriately to the feelings of others, he lost his job and went bankrupt. His marriage collapsed. And, at last report, he was dependent on custodial care and a permanent disability check.

In its elevation of rational, scientific thinking, the Enlightenment focused its attention on one branch of human intelligence: logical, deductive, proof-oriented mental operations. That branch of intelligence has brought us the scientific revolution. But a focus on rational, analytical intelligence fails to credit the importance of intuition in the creative process. Indeed, the scientific revolution was advanced by creative, intuitive insights which then had to be subjected to rigorous analysis and proof. John Maynard Keynes remarked of Isaac Newton that "His peculiar gift was the power of holding continuously in his mind a purely mental problem until he had seen straight through it. I [believe] Newton's preeminence is due to his [powers] of intuition being the strongest and most enduring with which a man has ever been gifted." Other examples of creativity and intuition abound. Archimedes stepping into his bath and exclaiming "Eureka!" Einstein reporting that he was guided by intuition and that "Words and language . . . do not seem to play any part in my thought process." In a recent survey, 72 out of 83 Nobel Laureates in science and medicine identified intuition as the cause of their success. Michael Brown, winner of the 1985 Nobel prize for medicine stated, "We felt at times there was almost a hand guiding us. We would go from one step to the next, and somehow we would know which was the right way to go, and I really can't tell you how we knew that."

The foregoing examples can be instructive in telling us how intuition can be most effectively employed. A crucial factor is experience, or, put another way, the development of expertise. A novice chess player, even if an intuitive genius, has not developed the experience and expertise to allow his intuition to play a meaningful role in playing the game. Gary Kasparov, on the other hand, because of his extraordinary expertise coupled with his intuitive genius, was able to beat IBM's Deep Blue computer, which was programmed with thousands of classic chess games and was able to calculate 200 million moves per second. Michael Jordan's intuitive genius (and that may be redundant since all geniuses are high intuitive) was best seen after he had developed his physical skills to the highest level. The same could be said for Tiger Woods and other professional athletes. In the world of business and finance we see people like Warren Buffett, John Templeton, George Soros and Jack Welsh applying their formidable intuitive judgments, but only after they had achieved a level of skill and experience, gathered all relevant information and undertaken careful and rigorous analysis. So intuition can best be employed when a foundation of expertise has first been established. As Louis Pasteur said "Chance favors only the prepared mind."

The flash of insight that can come to the prepared mind is well illustrated in a story involving Andrew Wiles, a professor of mathematics at Princeton. Wiles, together with most mathematicians throughout the world, had been fascinated by a famous challenge issued by Pierre de Fermat, a 17th century mathematical genius. In Fermat's so-called last theorem, he jotted in his own hand in a book alongside the Pythagorean theorem, a note stating that he had solved this equation (A squared plus B squared = C squared) for all exponents of any whole number greater than 2. His notes remarked "I have a truly marvelous demonstration of this proposition, which this margin is too narrow to contain."

For more than three centuries this puzzle has baffled the greatest mathematical minds in the world. Andrew Wiles had considered this problem for more than 30 years, coming back to it again and again. Then, one morning in June 1993, like a bolt out of the blue, the solution struck him. He stated "It was so indescribably beautiful; it was so simple and so elegant. I couldn't understand how I had missed it, and I just stared at it in disbelief for 20 minutes. Then, during the day, I walked around the department, and I'd keep coming back to my desk looking to see if it was still there. It was still there. I couldn't contain myself, I was so excited. It was the most important moment of my working life."

But for all of its powers, intuition has a perilous or dark side. For the truth is that intuition often offers a false or misperceived view of reality. As such, it misinforms our conscious minds and thus causes us to make mistakes. Let me cite a few examples. Take a look at the sheet of paper I have provided to each of you. Your intuition may tell you one thing about each of these questions, but chances are your intuition is wrong.

The dot is exactly half way up the triangle, though your intuition – your direct, nonanalytical knowledge, would say it's higher.

The two box tops are identical in size and shape, hard as that may be to believe.

The line segment A-B is not only longer than B-C, it is one-third longer.

The row houses in question D are of identical length, though our intuitions would continue to tell us that the row on the right must be shorter.

Finally, you are probably not familiar with the phrase "A bird in the the hand."

There are, in addition to what we have just discussed, many other ways in which our intuition fails to guide us correctly. For example, our own emotional states, our moods, can influence our judgments. When our emotions mix with our intuitions, we tend to perceive things in different ways. For example, if I am in a bad mood I may read someone's neutral look as a hostile glare, but if I am in a good mood, I might intuitively view that same look as one of interest and concern.

And, of course, our memories can infiltrate our intuitions and, as trial lawyers will confirm again and again, our memories are often quite flawed. Indeed, it is very likely that memories themselves are intuitions. Remember the definition of intuition as a direct, immediate apprehension without any rational analysis. Under this definition, memories certainly seem like they would constitute intuitions. If that is the case, then a flawed memory can result in a flawed intuition.

A third way intuition often leads us astray is in the manner in which we view ourselves. Studies over the years demonstrate convincingly that almost all of us see ourselves as better than the norm in nearly every area of endeavor. As Dave Barry, who is a very funny feature writer for the Miami Herald, once said, "The one thing that unites all human beings, regardless of age, gender, religion, economic status or ethnic background, is that deep down inside, we all believe that we are above average drivers."

And just like Lake Woebegon, where "all the women are strong, all the men are goodlooking, and all the children are above average," we all tend to see ourselves as more intelligent, better looking, less biased and more ethical than our peers. In a recent Gallup Poll, 44% of white Americans rated other whites as having a high prejudice against blacks.

Yet only 14% rated themselves as similarly prejudiced.

This elevated self-perception was also demonstrated in a question posed by U. S. *News and World Report* in 1997 about who was at least "somewhat likely" to go to heaven. 19% thought that O. J. Simpson would make the trip. They were more optimistic about Bill Clinton (52%), and Michael Jordan (65%). The second closest person to be perceived as a shooin was Mother Teresa at 79%. And who do you suppose topped Mother Teresa? At the head of the class, with an 87% admission rate, people placed themselves.

My overview of the perils of intuition must be a limited one. Psychologists have identified at least a dozen major ways in which our intuitions fail to guide us properly, including faulty memory, misreading our own thoughts or feelings, overconfidence, and deep-seated biases. The important point here is that since we are guided through much of our lives by unconscious intuitive thinking, although that intuitive thinking is powerful, quick and often quite persuasive to our conscious rational minds, it is often incorrect. So if we want to think smarter, we need to check our intuitions, whether we call them hunches, gut feelings or inner voices, against available evidence whenever we can.

One obvious way that individuals find themselves hijacked by mistaken intuition is in the area of gambling. Today Americans spend at casinos, lotteries, race tracks and the like more than $500 billion per year. That's up 30 times from the $17 billion spent in 1974. Gamblers leave about $50 billion behind, or 10% of all monies gambled. Why do these people gamble? Many because of the thrill they receive, and greed is probably another factor. But one of the largest motivators appears to be a total misunderstanding of the probabilities that apply in any gambling activity. Even though the evidence is plain that the probabilities of winning are not only small, but in some gambling activities such as the Powerball lottery, infinitesimal, people tend consistently to overestimate the opportunity of a favorable outcome for themselves. Their intuitions tell them that the law of probabilities doesn't necessarily apply to them.

One wonders whether some sort of warning should accompany a gambler's decision to bet. Like warnings accompanying packs of cigarettes, perhaps gamblers could be given a wakeup call about the reality of probabilities. For example:

- If you drive 10 miles to purchase a Powerball ticket, you are 16 times more likely to die in a car crash en route than to win the jackpot.

- If you play the Powerball lottery by purchasing one ticket per week, you would have to play for 1.6 million years to have a reasonable chance to win.

Finally, let's take a brief look at psychic intuition. A large part of our population is fascinated by the fact that humans may have the capacity to read minds, communicate with the dead, or predict the future. In a May 2001 Gallup survey, 50% of Americans declared their belief in extrasensory perception and only 27% said they didn't believe, with the rest being unsure. Recent TV shows like the X-Files, as well as movies such as The Sixth Sense, create powerful suggestions that ESP is a genuine phenomenon. As more confirmation, the Dial-A-Psychic industry topped $1 billion per year recently.

So what is the scientific response to the existence of ESP, whether it's characterized as clairvoyance, psychokinesis, telepathy, or precognition? The answer is that science has not found any credible evidence that these powers exist in any human being. After thousands of experiments, no ESP phenomenon has been found and no individual has been identified who can consistently demonstrate psychic ability. When I was at Duke, a professor, Dr. J. B. Rhine, who termed himself a parapsychologist, conducted numerous studies of extrasensory perception. For a time claimed that he had identified certain phenomenal individuals who, in turning over playing cards, could predict the outcome to an extent well above what would be chance, or 50/50. However, retesting these individuals months or years later, their precognition regressed to the norm (i.e. went back to 50/50) and their previous achievements appeared to be more like lucky streaks than any power of precognition.

But a determination by the scientific community that no credible evidence of extrasensory perception exists has not discouraged many people who want to believe in psychic powers. And then there are the many self-proclaimed psychics who have used this fascination with psychic intuition to separate gullible people from their money. Whether it is fortune telling, crystal ball reading, palmistry, tarot cards, mind-reading or psychokinesis, or so-called mind over matter, people want to believe that these powers in fact exist. One commentator who was critical of the mind over matter, or psychokinesis, claims of several individuals commented this way, "Will all those who believe in psychokinesis please raise my hand?"

Is there an explanation for why people continue to believe in ESP despite scientific conclusions to the contrary? Part of the explanation is probably the power of coincidence. The fact is that weird coincidences do happen, and when they happen they capture our attention and stay in our memories. We can all recall what seemed to be the remarkable coincidences that came out of the Kennedy assassination when comparisons were made with Lincoln's assassination 100 years before. You will recall they were each assassinated on a Friday, while seated beside their wives. Their assassins had the same number of letters in their names, and so on. And recall the fact that John Adams and Thomas Jefferson died on the same day, which just happened to be July 4, 1826, exactly 50 years the after signing of the Declaration of Independence. The coincidences of that event had a remarkable effect on Americans, many of whom concluded that God's hand surely was guiding the United States of America. I suppose the public's fascination with these kinds of coincidences and claims of extrasensory perception is a reflection of our innate desires to experience something that's transcendent or magical, and so we allow these desires to overwhelm our common sense and rational minds.

So we see that despite intuition's enormous powers, left unchecked by critical analysis and a responsible view of reality, intuition can surely lead us astray. And yet we must rely on intuition for much of our actions and judgments day in and day out. The key is not to be hijacked by unfettered intuition. As the physicist, Richard Feynman, said, "The first principle is that you must not fool yourself – and you are the easiest person to fool."

Sources:

Intuition by David G. Myers (Yale Univ. Press 2002)

Blink by Malcolm Gladwell (Little Brown and Co. 2005)

a. How far up this triangle is the dot?

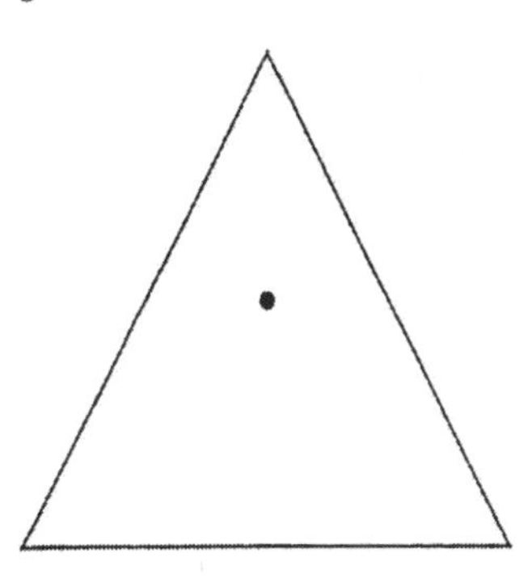

b. Do the dimensions of these two box tops differ?

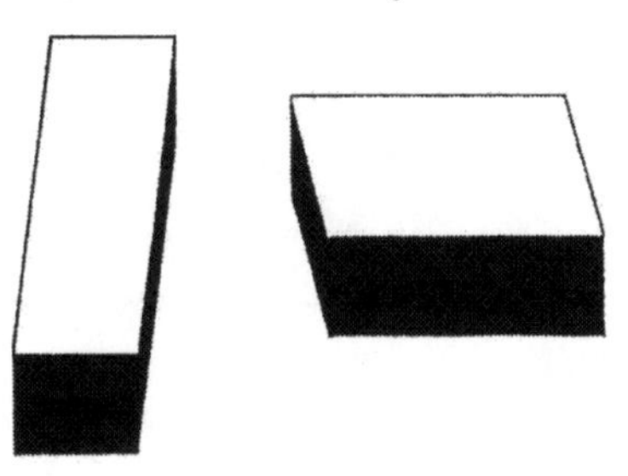

c. Which of these two line segments (AB or BC) is longer?

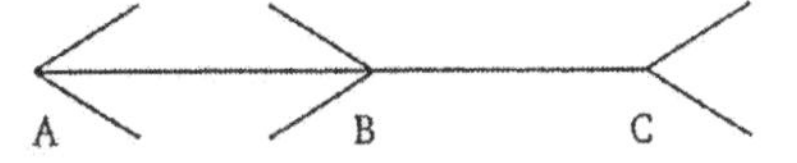

d. Line CD is what percent as long as AB?

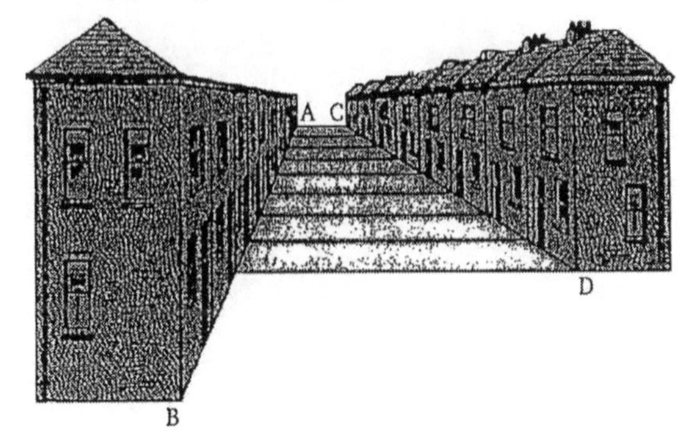

e. Are you familiar with this phrase?

A
BIRD
IN THE
THE HAND

The Round Table Presentation | February 7, 2008

How The West Was Won, Perhaps
The Challenge Of Radical Islam

In the year 732 AD a large Muslim army met a Frankish force near Tours, less than 200 miles south of Paris. In 711 a Muslim army had first crossed from Morocco into Spain and by 718 had conquered the entire Iberian peninsula. Now, only 14 years later, the Saracen general, Abdul Rahman, sought to bring France and the rest of Europe under Muslim dominion. In a battle that lasted 7 days, the army of the Franks defeated the Muslims who promptly retreated back into Spain. The victory earned the Frankish chieftan, Charles, the surname "Martel" or the "Hammer." Edward Gibbon in his "Decline and Fall" concluded that this battle was one of the most decisive in world history, for it determined whether Europe would fall under the sway of Islam or Christianity.

But now, 14 centuries later, Europe is confronted with the same prospect that faced Charles Martel albeit in a different context. Since World War II Muslim immigrants have flooded into western Europe. Do these immigrants represent the core of a revived effort to Islamify a significant part of Western civilization? A growing number of influential people think so. For example, Oriana Fallaci, a renowned Italian journalist, authored a book in 2004 entitled "The Force of Reason." Her thesis is that Europe is on the verge of becoming a dominion of Islam, and that the people of the

West have surrendered themselves meekly to the "Sons of Allah." In a Wall Street Journal interview in 2005 Ms. Fallaci stated that, "Europe is no longer Europe, it is Eurabia, a colony of Islam, where the Islamic invasion does not proceed only in a physical sense, but also in a mental and cultural sense. Servility to the invaders has poisoned democracy, with obvious consequences for freedom of thought and for the concept of liberty."

Strong words, and because of them, Ms. Fallaci faces jail in Italy for her beliefs. In a charge brought by a Muslim Italian citizen, Ms. Fallaci was indicted under a provision of the Italian penal code which proscribes the vilification of any religion permitted by Italian law, in this case, Islam. Other Muslims in Italy have publicly called for her execution.

Many of you may have already studied this subject in some depth. John Compton gave a thoughtful paper to The Round Table in 2003 entitled "Seeking Islam." My investigations and readings over the last several months have led me to be quite concerned about the ability of a militant strain of Islam to coexist with the free societies and democratic ideals espoused by the West. How will Europe and indeed the rest of the Western world address the issues of a religion that has grown increasingly radical and aggressive with its adherents urging Jihad against all things Western? Is militant Islam essentially incompatible with Western culture and ideals?

In addressing these questions, I thought I would begin by exploring with you a brief history of Islam, focusing especially on developments during the 20th century, to see how the West is confronting, or failing to confront, radical Islam.

Born in Mecca in 570 AD, Muhammad first became a merchant actively traveling with trading caravans throughout the Middle East. In 610 at the age of 40, Muhammad retired to meditate in a cave located on Mt. Hira just outside the city of Mecca. While meditating, he heard a divine voice, which he believed to be the angel Gabriel from the Jewish and Christian scriptures. The voice told him that his existing polytheistic notion of the divine was wrong and that Allah was the one God, that Muhammad must adopt the name of "Prophet," and that he must convert the tribes of Arabia to accept this new religion.

Over the next 12 years Muhammad continued to receive numerous revelations from Allah through Gabriel which now form the 114 chapters of the Koran. However, Muhammad's early efforts to convert the polytheistic tribes in and around Mecca were resisted, so he and his followers migrated to the town of Yathrib which was open to Muhammad's new faith. This migration is known as the Hijira and marks the formal beginning of Islam. Muhammad became the ruler of Yathrib and changed the name of the town to Medina, which means "city of the prophet." For the next ten years Muhammad and his followers constituted themselves as an army and began a series of conquests resulting in the subjugation of much of the Arabian peninsula, including Mecca, which he then declared to be the center of Islam.

Muhammad died in 632 and was succeeded by a series of four caliphs, or "successors," known in Islam as the "rightly-guided caliphs." The first successor, Abu-Bakr, was Muhammad's fatherinlaw. He ruled for two years and was succeeded by Umar who led his army to successful conquests of Syria, Palestine, and Persia. However, in 644 Caliph Umar was murdered and was succeeded by Uthman. Under Uthman, the armies of Islam moved into Egypt and began the spread of Islam through all of North Africa. Uthman was also responsible for bringing the diverse texts of the Koran together and creating one single consistent textual version of the Koran which has remained unchanged from that day forward. Interestingly, both Umar and Uthman established a precedent of permitting religious freedom for Christians, Jews and other religions in all the territories they conquered. So long as these non-Muslims paid a tax to their

Islamic overlords, thus rendering them effectively second-class citizens, the Muslims did not interfere with their daily lives or their freedom of worship.

Despite Uthman's military and political successes, because Uthman's family had initially rejected Muhammad's prophecies, many Muslim adherents objected to his caliphate, and in 656 Uthman was murdered by members of that group. Those who opposed Uthman then installed Ali as the next caliph; however, five years later Uthman's followers murdered Ali. The split between followers of Uthman and Ali became the most significant division within Islam. Ali's followers formed a religious party and called themselves Shiites. Ali was a cousin of Muhammad, and one of the first principles of the Shiite faction is that only descendants of Muhammad through Ali deserve the title of Caliph or have any right to exert any authority over Muslims. The opposing group, supporters of Uthman, became known as Sunnis. It was the Sunnis' position that the customs of the community should govern Islam and that the Caliphate should evolve by community consent rather than through a hereditary descent of spiritual authority.

By the time of Ali's death in 651 AD, Islam, through Muhammad and his 4 successors, the so-called "rightly-guided caliphs," in less than 30 years and solely by way of military conquest, had established dominion over an area stretching from the border of present India in the east to the Caspian Sea in the north, to eastern Turkey, the Arabian peninsula, and to all of current Syria, Jordan, Palestine, Egypt and much of North Africa.

Over the next several centuries there followed a flowering of science and literature throughout the Islamic world. But with the exception of the Ottoman empire that surged to power in the 1400's, the initial impulse of energy and zeal for conquest that caused the rapid spread of Islam in the 7th century had become quiescent, and Islam as a political force went into decline until the 20th century.

From our vantage point at the beginning of the 21st century, we can look back and see that the 20th century experienced a significant revival of a militant brand of Islam, an Islam that existed originally in the 7th century. While there may exist a number of reasons why this is so, I believe that there are at least three significant explanations for this potent rebirth of radical Islam. The first of these explanations has its origin in an Islamic cleric who lived during the 1700s but whose ideology has had an extraordinary effect on modern day Islam. The cleric's name was Mohammed Ibn Abdal Wahhab. Al Wahhab preached a puritanical, fundamentalist form of Islam which he characterized as a "purified" form of the faith. Al Wahhab's teachings at the time were controversial and were rejected by most of his contemporaries. Wahhabism urged Muslims to revert to the primitive, fundamental religious practices initiated by Muhammad in the 7th century.

In the single written work Al Wahhab produced, he gave specific instructions on the extension of Islam by force, just as Mohammed and his four rightly guided caliphs had done. All unbelievers, that is, Christians, Jews, Hindus, as well as Muslims who did not accept Al Wahhab's teaching, were to be put to death. Immediate entrance into paradise was promised to Al Wahhab's soldiers who fell in battle. Muslims must institute Sharia - Islamic government - throughout the world.

This extremely aggressive strain of Islam may well have been consigned to oblivion had not an alliance occurred between Al Wahhab and the leader of the Saudi tribe in Arabia, Ibn Sa'ud. When the Saudi dynasty took control of Saudi Arabia following World War I, Wahhabism became the official, indeed the exclusive, form of Islam practiced in Saudi Arabia. Osama Bin Laden is an adherent of Wahhabism. Al Qaeda has adopted its basic tenets. Moreover, as Saudi Arabia gained its extraordinary wealth from its oil exports, much of this wealth has been used to export Wahhabism throughout the world, including the United States, by establishing schools, called Madrassas, where Wahhabism is taught.

In 2005, a 23-year old American citizen named Ahmed Omar Abu Ali, was charged with plotting to assassinate President Bush. According to the New York Times, Ali was born in Houston and moved to Falls Church, Virginia where he was valedictorian of his high school class. Without more, it seems inconceivable that someone who had been so successful in high school could be charged with such a crime. But the high school he attended was actually a Madrassa, known as the Islamic Saudi Academy and funded by the kingdom of Saudi Arabia. And what does it teach? Wahhabi history and doctrine, including the demand that all who practice Wahhabism undertake Jihad and that Jihad requires death to infidels.

While Wahhabism was taking solid root in Saudi Arabia in the early 20th century and being spread throughout the Muslim and non-Muslim worlds, the second contributor to the revived radical Islam came into existence. This was an organization that called itself the Muslim Brotherhood, founded in Egypt in the 1920's by a 22-year old elementary school teacher, Hasan Al-Banna. The Muslim Brotherhood also adopted many of the tenets of Wahhabism, and at its core was the following statement of purpose: "Allah is our objective. The Prophet is our leader. Koran is our law. Jihad is our way. Dying in the way of Allah is our highest hope."

The Muslim Brotherhood grew to be a potent force in Egyptian politics and expressed its hatred for the state of Egyptian political affairs by assassinating its Prime Minister in 1948 and making no fewer than five attempts to assassinate Egyptian President Gamal Abdal Nasser. During the 1930's and 1940's, the Muslim Brotherhood became closely tied to the Nazis. Indeed Hitler's Germany provided financial support to the Muslim Brotherhood to agitate against the Jews in the

Middle East and, as a result, the Muslim Brotherhood became vehemently anti-Semitic. It was responsible for widespread publication of Mein Kampf and the Protocols of the Elders of Zion throughout the Arab world during this time. For those of you who may not be familiar with the Protocols of the Elders of Zion, it was a book created in Russia for the Tsar's secret police at the turn of the century, blaming the Jews for the nation's ills and claiming that a secret Jewish cabal was plotting to take over the world. Hitler used the Protocols to justify his "final solution."

Armed cells of the Muslim Brotherhood now exist throughout the Arab world. And while Egypt has taken strong efforts to quell the Muslim Brotherhood's activities there primarily by imprisoning thousands of its members, the Brotherhood has retained its identity and its wholehearted support for Jihad and terrorism. Among its many pronouncements is the statement that "Muslims must understand that their work in America is a kind of grand Jihad to eliminate and destroy the western civilization from within and sabotaging its miserable house by their own hands so that Islam is made victorious over all other religions."

While the Nazis supported the virulent anti-Semitism of the Muslim Brotherhood, Hitler also found a true believer in Mohammed Al Husseini, the Grand Mufti of Jerusalem. Al Husseini is the third reason for the spread of this new and ferocious militant ideology. As Grand Mufti, Al Husseini presided as the Imam of the Al Aqsa Mosque in Jerusalem, the highest Muslim authority in Palestine. From his earliest days he espoused a complete hatred for all Jews, and on two occasions in the 1920's incited anti-Jewish riots by claiming that Jews were plotting to destroy his Mosque. Al Husseini found the Nazis to be a strong ideological match with his anti-Jewish brand of Islam, and he met and schemed with Hitler and the Nazi hierarchy to create a proNazi Pan-Arabic form of government in the Middle East. During much of World War II Al Husseini lived in Berlin and was provided with accommodations by the Nazis, as well as a generous monthly stipend. In return he regularly appeared on German radio pronouncing Jews as the bitterest enemies of Muslims, and he implored the adoption of a Nazi final solution by all Arabs.

Husseini actively recruited Bosnian Muslims for the Waffen SS and personally oversaw the creation of two SS divisions, one of which was responsible for the extermination of as many as 90% of all Jews living in Yugoslavia. The Mufti became upset with Heinrich Himmler in 1943 when Himmler, through the Red Cross, sought to trade 5,000 Jewish children for 20,000 German prisoners of war. Himmler came around to the Mufti's thinking and the children were gassed. Prior to Husseini's death in 1974 he played a central role in the creation of the Palestinian Liberation Organization (PLO). Husseini's nephew, Rahman Abdul Rauf El Husseini, served his uncle and was a major player in Palestinian terrorism for nearly 40 years. You would recognize this nephew by the secular name he adopted in 1952, Yasser Arafat.

Interestingly, the virulent anti-Semitism that marks much of the Muslim consciousness in the 20th century is a relatively new phenomenon. Most historians agree that until the 20th century Jews and Muslims were able to coexist in relative harmony, and so this rise in anti-Semitism in the 20th century marks a break with the historical treatment of Jews by Muslims.

So now that we have reviewed three of the seeds that have sprouted into a militant, radicalized Islam in the 20th century, just how vulnerable is the West to being overtaken by this aggressive Muslim ideology? Part of the answer would appear to lie in demographics; part depends on the will of Western societies to address the challenge of militant Islam; and part depends on whether a strategy can be devised that will preserve the ideals of Western civilization and allow a peaceful coexistence with Islam.

Let's begin with demographics. Statisticians agree that a fertility rate of 2.1 live births per woman in a society is required to sustain the population of that society. The 2005 fertility rate for the United States was slightly better than 2.1. However fertility rates in Europe tell a different tale. Seventeen European nations are now at what demographers call the lowest low fertility rate, that is 1.3 births per woman. In theory, if their fertility rates are not increased, those 17 nations will find their population cut in half every 35 to 40 years. The three Mediterranean nations in Europe are now below the so-called lowest low fertility rate: Spain at 1.2 births per woman, Italy and Greece at about the same. And Russia truly has become the proverbial sick man of Europe. Not only is the life expectancy for males in Russia now at 58 years, but the fertility rate in Russia is at 1.1. At this rate Russia stands to lose half its population over the next 40 years.

By comparison Pakistan and Saudi Arabia have birth rates of 5.1 and 4.5 respectively. Somalia is at 6.8 births per woman, Afghanistan at 6.7, and tiny Yemen at 6.6. The average fertility rate for European Union nations is 1.35 births per woman. The average for all countries where the Islamic population is a majority is 3.5.

And I might add that while the United States is at population stability, our northern neighbor, Canada, is at 1.4 live births per woman.

Moreover, in Europe there is evidence that an inordinate percentage of the births are occurring to Muslim women. Currently there are officially approximately 20 million Muslims living throughout the countries of the EU (though many observers believe there could be twice that many), and their fertility rates appear to be consistent with the average for Islamic countries in the rest of the world. That means that European women are experiencing fertility rates that are even lower than the published rates. In France, for example, where at least 10% (7 million) of the population is Muslim, approximately one-third of all live births are born to Muslim parents. Over the last several years the most popular boy's name in Belgium, Holland and Sweden has been Mohammed.

So we have an aging European population that for the most part is living in nations that provide significant welfare benefits to their citizens. How can these welfare benefits be sustained in countries where the population of those who are working to support the economy is dwindling so significantly? The answer, of course, is immigration. Following World War II Europe experienced an enormous influx of immigrants, primarily from Muslim countries. Brought in initially as "guest workers" to help rebuild war-ravaged Europe, European nations have become dependent upon immigration to maintain the vibrancy of their economies. And most of those immigrants have moved to the larger cities of Europe. It is estimated that 1 million Muslims live in each of London and Paris. The Muslims in Amsterdam and Rotterdam make up approximately 40% of the population of each of those cities, and Scandinavia has seen a huge influx in Muslim immigrants over the past decade. Malmo, Sweden, a city of 250,000, is now more than 25% Muslim. Recent pronouncements from the European Union suggests that over the next 25 years or so Europe will need an influx of as many as 50,000,000 new immigrants in order to maintain a large enough working population to fund the social programs that are currently in place. Political commentators in various European countries appear to have accepted significant immigration as absolutely necessary to maintain the standard of living to which they have become accustomed. An editorial in the Toronto Globe and Mail echoes the sentiments of various commentators in Europe. Reacting to the lowest Canadian fertility rate since records began and a 25% fall since 1992, the Globe and Mail wrote: "Luckily for our future economic and fiscal wellbeing, Canada is wellpositioned to counter the declining population trend by continuing to encourage the immigration of talented people to this country from overcrowded parts of the world." It turns out that the second most populous immigrant group moving to Canada is from Muslim countries.

"We are the ones who will change you," Norwegian Imam Mullah Krekar told an Oslo newspaper in 2006. "Just look at the development within Europe, where the number of Muslims is expanding like mosquitoes. Every Western woman in the EU is producing an average of 1.4 children. Every Muslim woman in the same countries is producing 3.5 children." As he summed it up: "Our way of thinking will prove more powerful than yours." It's not just their way of thinking but their way of procreating that will prove more powerful than that of the Europeans.

And Colonel Muammar Gaddafi made the following recent statement: "There are signs that Allah will grant Islam victory in Europe – without swords, without guns, without conquests. The 50,000,000 Muslims of Europe will turn it into a Muslim continent within a few decades."

Implicit in the foregoing comments is the notion that Muslim populations in Europe intend to maintain their identity, to avoid assimilation, and will strive to replace the culture and democracies of Western Europe with societies controlled by Islamists and governed by Sharia (Islamic law). And one cannot help feeling that the evidence supports the notion that Muslims in Europe have the religious and political will to carry out the desired transformation of Europe and perhaps the rest of the West into a dominion of Islam. For Islam is not just a religion. It is also a legal code, a guide for personal behavior, and a commitment to Jihad against the infidel. As one commentator stated, "The West is dealing with a self-segregating religion housing a global terrorist movement supported by its holiest book and whose adherents are the fastest growing demographic in the developed word." It is an ideology that uses its houses of worship, its mosques, as recruiting centers for Jihad. Many of the mosques have clear links to terrorist cells throughout the world, and even the more so-called moderate Imams in Europe and the West support the ultimate objectives of the terrorists, even if they have no direct link to the terrorists themselves.

What also seems clear is that the European democracies do not appear to be adjusting successfully to this infusion of radicalized Muslims. Europeans and Muslim minorities are espousing clearly different priorities and objectives. For the Europeans, it is health care, retirement, social programs, a commitment to multiculturalism and diversity, a primarily secular world view, and a reverence for life. Muslims in Europe are committed to a world view consistent with their counterparts in Islamic countries, a world view that sees a single world religion – Islam, a society whose civil institutions are guided by the Koran, a commitment to Jihad and Sharia, a rejection of diversity, a demand for uniformity, and a culture where death in the name of Allah is to be sought after. If we characterize this confrontation between Islamic immigrants and their European hosts as a battle of wills, it would appear that the Islamists currently have the upper hand.

In addition to the horrendous acts of terrorism that have occurred in London, Madrid, Beslan, Paris and on the streets of Amsterdam, there is every evidence that militant Islam has embarked on a more subtle strategy, one that does not attack European civilization head on, but works in a way to make Europeans uncomfortable, to disrupt their lives at the margin, hoping that they will look for the easiest path to renormalization. And, as a result, it appears that each European host country is working hard to assimilate itself with the militant Islamists living there rather than the other way round.

French municipal swimming pools have introduced gendersegregated bathing sessions at the request of Muslim activists. Burger King withdrew ice cream cones from its British menus because a young Muslim activist in London complained that the creamy swirl shown on the lid of these ice cream cones looked like the word "Allah" in Arabic script. I am sure you remember the socalled "cartoon Jihad" of early 2006. It involved the publication by one Danish newspaper of various cartoonists' representations of the prophet Mohammed. The result was weeks of protests, lawsuits, death threats, rioting and killing. Not only in Europe but across the globe. Muslim protestors outside the Danish Embassy in London carried placards that read "Exterminate those who slander Islam, behead those who insult Islam, Europe you will come crawling when the Mujahideen come roaring." One result of these protests was that the person heading the Justice and Security Commission for the EU declared that Europe would set up a "media code" to encourage prudence in the way certain sensitive subjects are covered. He went on to say, "We are aware of the consequences of exercising the right of free expression . . . . we can and we are ready to self-regulate that right."

In Austria Muslims, who make up approximately 5% of the population, are demanding that all female teachers, believers or infidels, wear head scarves in class. Women in France, Holland and Belgium have elected to be "covered" when in public to avoid the criticism of Muslims whom they meet in the streets. The Muslim Counsel of Britain wants Holocaust Day abolished because it focuses "only" on the Nazis holocaust of the Jews and not the Israelis on-going holocaust of the Palestinians. In Seville, King Ferdinand III is no longer patron saint of the annual fiesta because his splendid record in fighting for Spanish independence from the Moors was felt to be insensitive to Muslims. In London, a judge agreed to the removal of Jews and Hindus from a trial jury because the Muslim defendant's counsel argued he couldn't get a fair verdict from them. The Church of England is considering removing St. George as the country's patron saint on the ground that, according to certain Anglican clergy, he is too militaristic and is offensive to Muslims. In 2005, the Chief Inspector of Prisons in the UK banned the flying of the English National Flag in English prisons on the grounds that it shows the cross of St. George, which was used by the Crusaders and so is offensive to Muslims.

You may have heard of Abu Hamza, who was the former Imam of Finsbury Park Mosque in London. On trial in London for nine counts of soliciting murder his distinguished Queen's counsel Edward Fitzgerald, QC, raised the defense that Hamza's interpretation of the Koran was that it imposed an obligation on Muslims to do Jihad and to fight in the defense of their religion. According to the counsel Hamza was not preaching murder, he was simply preaching from the Koran.

And the examples go on and on, even in the United States. Just before Christmas 2003 Muslim leaders in California applauded the decision of the Catholic High School in San Juan, Capistrano, to change the name of its football team from the Crusaders to the less culturally insensitive Lions. Meanwhile, just 20 miles away in Irvine, the four Muslim football teams in Orange County retained their names: the Intifada, the Mujahideen, the Saracens, and the Sword of Allah.

Then there's the case of Abdurahman Alamoudi who was jailed in 2003 for laundering money from a Libyan terror front "charity" into Syria. Alamoudi is the person who until 1998 certified Muslim chaplains for the United States military. He is also the person who helped devise a three week Islamic awareness course for the California public schools in the course of which students adopt Muslim names, wear Islamic garb, give up candy and television for Ramadan, memorize suras from the Koran, learn that Jihad means struggle, profess the Muslim faith, and recite prayers that begin in the name of Allah. The Ninth Circuit Court of Appeals, the same court that ruled the pledge of allegiance unconstitutional because of the words "under God" decided in this case that allowing seventh graders to experience aspects of Islam for a twoweek period was acceptable, an exposure to a fascinating culture from which every pupil can benefit.

Assuming that the West is able to understand what it is confronting and is able to summon the will to actually confront militant Islam within its borders, what strategies are available to allow Western culture and radical Islam to coexist? We have already seen that assimilation is not likely. Is there a possibility that Islam itself can be reformed, if not from the outside, then internally by more moderate Muslims? We have heard since September 11 that moderate Muslims do not support acts of terrorism. They do not support calls for violence against innocent people. One of the problems I have had in my investigations is trying to determine just who are these moderate Muslims.

On the first anniversary of the July 7, 2005 London Tube bombings, the Times of London commissioned a poll of British Muslims. Some of the findings included the following:

- 16% say that while the attacks may have been wrong, the cause was right.
- 13% think the four men who carried out the bombings should be regarded as martyrs.
- 7% agree that suicide attacks on civilians in the United Kingdom can be justified in some circumstances.
- 2% would be proud if a family member decided to join al Qaeda.

If there are 1 million Muslims in London and 7% think that suicide attacks on civilians are justified, that's 70,000 potential Islamic extremists in Britain's capital city. Remember it only took 19 on September 11.

According to a different poll, a large majority of British

Muslims support nearly all of the terrorists' long-term strategic goals. Over 60% of British Muslims want to live under Sharia in the United Kingdom.

Or take the comments made by a self-proclaimed moderate Muslim woman who is a kindergarten teacher in Lansing, Michigan:

> *Islam is not only a religion, it is a complete way of life. Islam guides Muslims from birth to grave. The Koran and prophet Mohammed's words and practical application of Koran in life cannot be changed.*
>
> *Islam is a guide for humanity, for all times, until the day of judgment. It is forbidden in Islam to convert to any other religion. The penalty is death. There is no disagreement about it.*
>
> *Islam is being embraced by people of other faiths all the time. They should know they can embrace Islam, but cannot get out. This rule is not made by Muslims; it is the supreme law of God.*

This woman would agree with the Islamists in post-Taliban Afghanistan who put a man named Abdul Rahman on trial for his life because he had committed the crime of converting to Christianity. The leader of the primary Islamic religious body in Afghanistan said, "We will not allow God to be humiliated. This man must die. Cut off his head. We will call on the people to pull him into pieces so there's nothing left." The man who spoke these words was described as a leading moderate cleric in Afghanistan.

Here's another example. Souleiman Ghali was born in Palestine. After immigrating to America, he found himself rethinking the old prejudices against Shiites, Christians and Jews and in 1993 he helped found a mosque in San Francisco. Mr. Ghali's website stated, "Our vision is the emergence of an

American Muslim identity founded on compassion, respect, dignity and love." In 2002 Mr. Ghali fired an Imam who urged California Muslims to follow the example of Palestinian suicide bombers to bring Jihad to America. Safwat Morsy is an Egyptian and he speaks barely any English, but he knew enough to sue Mr. Ghali's mosque for wrongful dismissal and was awarded $400,000.

All that is not surprising, at least in California, but that's not the full story. Other moderate Imams who are on the Board of the Mosque sided with the proterrorist assertions of Safwat Morsy, so they forced Mr. Ghali off the Board and out of any role in the mosque that he founded. Mr. Ghali's views were simply too moderate for the Muslim leadership in San Francisco.

I am willing to believe that the term "moderate Muslim" is not an entirely fictional one. But I also am convinced that they are quiescent at the present time. Just like there were a lot of moderate Germans during the 1930's, moderate Muslims today bring little if anything to the table in terms of reforming militant Islam from within. Moreover, it appears that even if there are moderate Muslims, it may be an oxymoron to declare there is a "moderate Islam" today. Certainly millions of Muslims just want to get on with their lives, to raise their children and practice their religion, and there are certainly places in the world where Muslim practices have reached accommodation with different customs in a community, even Western customs. However, as far as I can tell all the official schools of Islamic jurisprudence, and the Koran itself, commend violent Jihad and the establishment of Sharia, that is, Islamic law governing an entire society. So even a "moderate Muslim" has difficulty in finding any formal authority to support an argument for moderation. And to be a moderate Muslim publicly means standing up to the leaders of your community, to men like Shaker Elsayed, leader of one of America's largest mosques, who has told his Islamic fundamentalist followers, "The call to reform Islam is an alien call." And of course we know examples of people who have spoken out against militant Islam. The Iranians declared a fatwa on Salman Rushdie, and he had to go into hiding for more than a decade. The Dutch film maker, Theo van Gogh, spoke out and was murdered. Ayaan Hirsy Ali, a Somalian woman and Muslim who was in the Dutch Parliament, and who spoke out against militant Islam, was forced to leave Holland and find protection in the United States against numerous death threats from offended Muslims for her urging of moderation. I note that John Compton's paper on Islam closed with a quote from a column by Thomas Friedman in the New York Times on December 4, 2002. In his column he expressed the need for what he termed "Islamic humanism" saying, "Just as the protestant movement wanted to rescue Christianity from the clergy and the church hierarchy, so [moderate] Muslims must do something similar today." I hope that can happen but I must admit I am not very hopeful that militant Islam presently can be reformed from within the Islamic community as a whole.

The verdict, of course, is still out. All of us in the West are slowly but steadily coming to a clearer understanding of Islam and especially its radicalized new form. We have a clearer view of the long-term strategy of militant Islam. We know that terrorism will be part of the Jihad against the non-Islamic world. But there are Muslim countries, such as Turkey, Egypt and Jordan, that espouse a peaceful coexistence with the West. Remember also that the initial spread of Islam in the 7th century made an early accommodation with Christians and other non-Muslims who fell under Islamic rule. Nonetheless, I do not yet see a clear answer as to how we find a resolution to this conflict. The Islamists living in Europe and in Canada and the US have used, even abused, our Western tradition of tolerance and acceptance of diversity to further their objectives. So far the West's position has been to try and accommodate Islam, even militant Islam; however, the Muslims for their part do not accommodate. So far it seems entirely one sided. I believe we must be concerned that Islamists will interpret the deference given them by people of good will in Western societies as indicative of a lack of any real conviction to stand firm for Western ideals. And I do believe that if we cannot summon a will to prevail against these inroads, we may ultimately lose the battle.

Sources:

Mark Steyn, *America Alone*, Regnery Publishing, Inc., 2006

Samuel Huntington, *The Clash of Civilizations*, Simon and Shuster, 1996

The Round Table Presentation | March 4, 2010

The Blackbirds
The Triumphant Tuskegee Airmen

This past January, as I was reading the New York Times, my eye was caught by an obituary for Lt. Col. Lee Archer, a World War II fighter pilot, who passed away at the age of 90.

Col. Archer flew 169 combat missions in the European theater of operations, flying out of Italy and escorting longrange bombers on raids over Nazi occupied Europe. In the process Archer shot down five enemy fighter planes. Col. Archer was black. He had been part of the first class of fighter pilots to graduate from Tuskegee Army Airfield in Tuskegee, Alabama, in 1943. He was one of the Tuskegee Airmen.

I had a vague recollection of the Tuskegee Airmen, and the obituary of Lee Archer sparked my curiosity, and so tonight, I would like to share with you briefly the story of the Tuskegee Airmen. As you may know, it is a story of how several thousand young black men volunteered to become pilots and ground crews for four all-black fighter squadrons and endured the segregationist policies and customs that dominated America and its institutions (especially the military) at the time. Despite the degrading treatment they faced, these young men formed the opening wedge that led to the desegregation of the American military. The story of the Tuskegee Airmen was a crucial step in America's agonizingly slow climb to full racial integration.

Blacks had been a part of the American military since the Revolutionary War. However, except in rare and unofficial situations, blacks were always part of segregated units. In World War I, black volunteers were allocated to "service" units, which included kitchen duties, cleaning, laundry, and other support services for the white troops. Following World War I, because of the pressure imposed primarily by the NAACP, studies were undertaken regarding the role of African Americans in the military. In 1925 the War Department undertook a study of the combat records of black servicemen during World War I. Because the segregationist policies in the American military restricted blacks to serve as stevedores, laborers and kitchen help during World War I, examining their "combat records" seemed to be singularly unhelpful. The study, entitled "The Use of Negro Manpower in War," gave vent to some extreme racial biases. It concluded that black men "were cowards and poor technicians and fighters, lacking initiative and resourcefulness." It reported that the brain of the average black man weighed only 35 ounces compared to 45 ounces for an average white man. The report concluded that African Americans were "a subspecies of the human population." This is language you would expect to see in some screed distributed by the Ku Klux Klan.

As Nazi aggression propelled Europe toward war, in April 1939 Congress passed a law creating a Civilian Pilot Training Program to provide private training of military pilots at civilian schools. At the last moment a Wyoming Senator added an amendment that allowed African American colleges to offer the Civilian Pilot Training Program. The U.S. Army Air Corps balked at the idea of training of black pilots and refused to submit to the legislative mandate. Congress responded in June 1939 by specifically directing the Civil Aeronautics Authority to sanction civilian pilot training at several black colleges, including Tuskegee Institute, and to furnish the aircraft for the program. Civilian pilot training for young black men began at Tuskegee in October 1939.

As these men completed their civilian pilot training, they sought to join the Army Air Corps. Every application made (and there were hundreds) was rejected based upon the single fact that the applicant was a Negro.

In the autumn of 1940, a Howard University student named Yancey Williams, filed a federal civil rights lawsuit based on the rejection of his numerous applications to become a flying cadet based solely on his race. The suit was brought against the Secretary of War, Henry Stimson, the Army Chief of Staff, George Marshall, and others. Thurgood Marshal, who was then a member of the NAACP legal staff, drafted and filed the complaint.

It so happened that in the fall of 1940, FDR was in an election contest against Republican Wendell Willkie. Willkie made a campaign promise that if elected he would end segregation in the military. This forced Roosevelt's hand, and prior to election day Roosevelt directed the War Department to issue a policy declaring that black men would be able to join the Army Air Corps. He made one stipulation, however; the men would be required to be part of a segregated, all-black flying unit.

So in January 1941, the War Department and the Army Air Corps, bowing to political pressure, announced that African Americans would be allowed to become pilots in the Air Corps, and it established the 99th Pursuit Squadron (Pursuit later became Fighter) to be activated and trained at Tuskegee Institute in Tuskegee, Alabama. The first class would begin with 33 to 35 pilots and 278 ground crew. The first officers would be white, but as black officers were trained, they would take command of the squadron. Thurgood Marshall dismissed the lawsuit, and the story of the Tuskegee Airmen was born.

The NAACP loudly objected to the plan to have a segregated unit. George Marshall responded to their complaint by offering a very carefully worded rejection of the demand for integration. He stated, "Segregation is an established American custom. The educational level of Negroes is below that of whites; the Army must utilize its personnel according to their capabilities; and experiments within the Army and the evolution of social problems are fraught with danger to efficiency, discipline and morale."

African American leaders at the time wanted the first all black fighter squadron to be located in the Chicago area. And, indeed, the first training of the black ground crews for the 99th Squadron was initiated in the spring of 1941 at an airfield in Chanute, Illinois, about 130 miles south of Chicago. But Tuskegee Institute was the choice, although it had no airfield that could accommodate the training that was required. So an immediate construction contract was let in the spring of 1941 for the construction of an airfield and barracks at Tuskegee Institute. The contract was awarded to McKissack and McKissack, an African American architectural firm located in Nashville, Tennessee, that many, if not most, of us know well and that is still in existence today.

There were some good reasons for selecting Tuskegee. The weather was better. The Tuskegee Institute was already part of a civilian pilot training program. Tuskegee, which had been founded by Booker T. Washington and had been in existence since 1881, was, along with Fisk and Howard Universities, one of the best known black colleges in the United States.

Shortly after the Army Air Corps announced its decision to create the 99th Pursuit Squadron at Tuskegee, Eleanor Roosevelt threw her support behind the project. She visited Tuskegee Institute in the spring of 1941. Charles "Chief" Anderson, an African American who headed the civilian pilot training program, invited her to take an airplane ride and Mrs. Roosevelt accepted, over the vehement objections of her Secret Service agents. One of the agents then called FDR who replied, "Well if she wants to do it, there is nothing I can do to stop her."

On July 19, 1941, eleven cadets and one officer were inducted into the 99th Pursuit Squadron, part of the first class of the Tuskegee Airmen. In August the 250 black enlisted men training at Chanute Field in Illinois were transferred to Tuskegee. The officer who was inducted into that first class was Capt. Benjamin O. Davis. In many ways the success of the Tuskegee experience was dependent on the leadership shown by Davis during the war.

In 1936 Benjamin Davis was the first African American to graduate from West Point in the 20th century. His years at West Point, from 1932 to 1936, were more difficult than any of us could imagine. As the lone Negro at the Academy, the cadre of other cadets refused to speak to him because of his skin color except when formally required to do so, and so Davis endured a veil of silence for his entire four years. Moreover, upper classmen issued a large number of undeserved demerits to Cadet Davis in a effort to cause his dismissal from the Corps. Fortunately, the Commandant at West Point, Simon Bolivar Buckner, realized that most of these were imaginary infractions and he voided those demerits and Davis was able to graduate. Despite what we would consider inhuman treatment by his peers, Davis performed well at West Point and graduated 35th in a class of 276 cadets.

Davis requested Air Corps pilot training after graduation in 1936. His application was denied in a letter that stated "there was no place in the Air Corps for a Negro pilot." But when the 99th Pursuit Squadron was finally approved by the War Department and the Air Corps in 1941, Davis was the first to apply.

Fortunately, although the flight instructors for this first class of Tuskegee Airmen were white, the cadets nonetheless considered those instructors to be fair. Bigotry, racism and mistreatment by the instructors did not appear to exist, and the black pilots felt that these white instructors genuinely wanted them to succeed.

Unfortunately, the white commanding officer at the Tuskegee Army Airfield did not. Col. Von Kimble enforced a rigorous segregation policy. Outside of the training program, socializing between whites and blacks on the base was absolutely prohibited. Eating areas and toilets were segregated.

The relationship between the African Americans on the base and whites in the Tuskegee community was also racially charged. Most restaurants were not available to blacks, and those that were had segregated areas. The clothing stores forbade the wives of the airmen from trying on clothing since white women refused to buy clothing that had been worn at any time by a black woman.

On a brighter side, the second commanding officer, Col. Parrish, relaxed the segregation policy to some degree. He realized the Tuskegee Airmen were a focus of pride for African Americans across the United States, and he welcomed black entertainers and celebrities who were eager to come to Tuskegee to meet and entertain the black troops there. Over the course of the several years of training at Tuskegee, the Airmen were visited and entertained by the likes of Cab Calloway, Ella Fitzgerald, Count Basie, Lena Horne, Rochester and Joe Lewis to name a few.

By April 1943, a total of 400 pilots and enlisted men who comprised the 99th Fighter Squadron were deemed ready for combat duty and were relocated to New York where they embarked for Casablanca. From Casablanca they moved in stages to Tunis where in June 1943 they actually began flight operations against the Nazis. The squadron commander was now Lt. Col. Benjamin Davis.

When the 99th arrived in Tunisia, it was met by Lt. Col. Philip C. Cochran, who was considered to be the best fighter pilot instructor in the entire Army Air Corps. Cochran trained with the pilots of the 99th for two weeks, simulating dogfights and focusing on tactical decisions to be made when up against enemy fighters. Cochran, who had a bigger than life personality, was the model for Flip Corkin in Milt Caniff's comic strip, Terry and the Pirates.

Initially provided with P-40 Warhawks, they were first involved in bombing the German held island of Pantelleria as part of the allied move from North Africa to Sicily. Accompanying B-17 bombers on a mission to Sicily, on July 2, 1943, Lt. Charles Hall of the 99th became the first Tuskegee Airman to shoot down a German aircraft, a Messerschmitt ME9. Following his return to base he received personal congratulations from General Eisenhower, and received the only bottle of Coca Cola in the squadron that had been saved for that occasion.

In September 1943, Lt. Col. Davis was reassigned back to Tuskegee to train new pilots and to become Commander of what was the all black 332nd Fighter Group. This group was comprised of three new fighter squadrons, the 100th, the 301st and the 302nd. The 99th would become part of the 332nd Fighter Group when it came to Italy in 1944 and, together, those four squadrons would comprise the Tuskegee Airmen.

Until the 332nd arrived, the 99th was assigned to the 33rd Fighter Group, consisting of all white fighter squadrons and commanded by a Col. Momyer. Unfortunately, Momyer had nothing but contempt for black aviators. He assigned the 99th shore patrol and bombing missions while their white counterparts were confronting the enemy in the skies over southern Italy. After the 99th joined his fighter group, he declared that “Negroes that were part of the 99th failed to display the aggressiveness and desire for combat that are necessary to a first class fighting organization.” He recommended that all black squadrons should be reassigned to noncombatant roles. His report was endorsed by General Edwin House, his immediate superior, and was sent to General Henry “Hap” Arnold, the Commanding General of the Army Air Force. Time Magazine reported on September 20, 1943 that “the question of the 99th is only a single phase of one of the Army’s biggest headaches: how to train and use Negro troops. No theater commander wants them in considerable numbers; the high command has trouble finding combat jobs for them. There is no lack of work to be done by Negroes as labor and engineering troops – the Army’s dirty work.” Time ended its article by asking “Is the Negro as good a soldier as the white man?”

To make matters worse, Lt. General Carl Spaatz, Deputy Commander of the Mediterranean Allied Airforces, added his opinion that the 99th be reassigned to coastal patrol duty in a location such as the Panama Canal zone.

The Time Magazine article and the comments of the generals created an enormous stir in the United States, and a committee of the War Department, the McCloy Committee, was promptly convened to consider the allegations. Lt. Col. Davis testified before that Committee, noting Col. Momyer had left the 99th some several hundred miles behind the active war zone in southern Italy. The 99th was left to fly missions over Sicily where they encountered few if any enemy aircraft. War correspondent Ernie Pyle had covered the 99th and publicly defended their performance, and General Eisenhower wrote to the Committee that he believed Momyer’s charges simply were not accurate.

General Hap Arnold then asked for a study to be made by the Pentagon looking at the performance of the 99th as compared to other squadrons. That study noted that the

99th had been given hand-me-down planes, that it had been given lower priority missions, but it also showed that the 99th Fighter Squadron was disciplined, committed, and as good or better than the other War Hawk units in the Mediterranean.

In the late summer and early fall of 1943, the 99th Fighter Squadron moved from Tunisia to southern Italy. Here the nature of their responsibilities changed. They had a new commanding officer with a new attitude. Instead of flying patrols far behind enemy lines, their new responsibility was to provide air support for the Salerno and Anzio beachheads. This meant the pilots of the 99th were constantly engaging enemy fighters. The Tuskegee Airmen flew as many as 50 sorties a day. On January 27 and 28, 1944 the 99th was sent up along with two squadrons commanded by Flip Cochran against a powerful German attack made up of over 100 Messerschmitts and Focke-Wulf aircraft. The 99ths' two-day total was impressive. They had twelve kills, three probables and four damaged enemy aircraft. They proved to be the most successful of the three squadrons that had gone up against the Luftwaffe.

Time Magazine, which had months earlier printed the derogatory article based on the Momyer Report, ran an article regarding the 99ths' successes on the 27th and 28th of January. Time reported, "Any outfit would have been proud of the combat excellence of one of the most controversial outfits in the Army . . . . They had finally got their big chance flying cover for the Allies beachhead and they knew what to do with it . . . . The squadron was veteran, wellled, sure of itself . . . . The Air Corps regards its experiment proven, and is taking all the qualified Negro cadets it can get." General Hap Arnold awarded the 99th a Distinguished Unit Citation. Intense aerial combat continued for the next several weeks and the 99th was in the thick of it as the Allies moved from the Anzio beachhead to the German stand at Monte Cassino.

As the 99th Fighter Squadron was performing with great success in Italy, the 332nd Fighter Group was preparing to join them. The group was composed of three new fighter squadrons and was commanded by Lt. Col. Benjamin Davis. It arrived in Taranto, Italy in February 1944. The 332nd was provided with P-39 Airacobras and for the next several months performed primarily coastal patrol missions and trained for aerial combat.

Then in July 1944, the 99th joined the three squadrons of the 332nd Fighter Group. They were equipped with new P-51 Mustangs, and their primary role now changed. The new responsibility for the Tuskegee Airmen was to serve as fighter escorts for bombing missions into the heart of Europe. The P-51 Mustang was well suited for long range escort missions. It had a greater range and speed, and with additional aluminum tanks that could be jettisoned when empty, could escort a bombing group as far as Berlin.

Upon receiving these new P-51 Mustangs, the pilots of the 332nd wanted to create a distinguishing feature on the planes. They found a large supply of bright red paint available at the base, and so they determined to paint the tail sections of the P-51s a bright red. For the rest of the war this all black fighter group was known as the Red Tails. Initially there was some concern that the red tails of the 332nd would make the planes an easier target for enemy fighters. That did not occur. Instead it made identification easier for the gunnery crews on the bombers, so there was less chance of damage to or loss of the P-51s as a result of friendly fire. And it also helped the fighter pilots in the thick of a dog fight to determine who was whom.

Between July 1944 and May 45, the Red Tails served as fighter escorts for hundreds of bombing raids into Yugoslavia, Greece, Bulgaria, Romania, Austria, France and Germany. The Red Tails earned a reputation of never having lost a bomber to enemy fighters during all those missions. That turned out to be a slight exaggeration, and after the war it was determined that perhaps as many as 25 bombers escorted by the Red Tails had been shot down by enemy aircraft. But the actual number will never be known because many bombers were lost to ground fire. What is true is that the reputation of the Red Tails in that final year of war caused many of the all-white bomber groups to specifically request to be escorted by the Red Tails. By all reports, the 332nd earned an impressive combat record during this time. During its escort of a group of B17 bombers to the Daimler-Benz factory in Berlin in March 1945, the Red Tails were engaged by 25 new German fighters including 11 new jet aircraft, and were credited with shooting down four jets and damaging seven other German fighters while losing three P-51s during the mission. The efforts of the Tuskegee Airmen on that day resulted in another Distinguished Unit Citation from the Army Air Corps. On March 31, 1945, escorting bombing missions, the Red Tails shot down 11 enemy planes. And on another redletter day, April 1, 1945, 8 Red Tails encountered 16 enemy fighters. They shot down 12 and damaged another two. Two of the Red Tails were lost. The Luftwaffe gave them the nickname "Schwarze Vogelmenschen" or black birdmen. The bomber pilots of the Fifteenth Air Force and crews called them the Red Tail Angels, and many of them proclaimed that the most beautiful things they ever saw in the skies of southern Europe were the red-tailed Mustangs of the Tuskegee Airmen.

By the end of the war the Tuskegee Airmen were credited with 112 Luftwaffe aircraft shot down, the sinking of a German destroyer by Red Tail machine gun fire, and the destruction of innumerable fuel dumps, trucks and trains. The squadrons of the 332nd Fighter Group flew more than 15,000 sorties on 1,500 missions, including 6,000 for the 99th prior to its joining the 332nd in July 1944.. On average a Tuskegee Airman flew twice the number of missions as their white counterparts during the war.

While the Tuskegee Airmen were distinguishing themselves in the European theater, a troubling event occurred in the United States that became known as the Freeman Field Mutiny, and while it is not about the Tuskegee Airmen, per se, it is part of the struggle of black America and the U. S. Military.

After the deployment of the 99th Fighter Squadron to Europe in 1943, and because of significant pressure exerted by African American leaders, newspapers, unions and civic groups, the Army Air Corps and the War Department reluctantly established an all black bombardment group. The 477th Bombardment Group was intended to have a compliment of 1200 Negro officers and men, and was allocated 60 B-25 twin engine Mitchell Bombers. This was a much larger project than the creation of a fighter squadron. The fighter plane needs only a pilot and a ground crew of two or three at the most. Bombers required a crew of 12, including pilots, navigators, bombardiers, gunners and ground crew.

The men of the 477th were eager to break down yet another military barrier to African Americans. Prior to the unit's reluctant activation, the Army Air Corps had issued a report on the formation of the group, concluding, "It is common knowledge that the colored race does not have the technical or flying background for the creation of a bombardment-type unit." And the actions of the Army Air Corps upon formation of the 477th indicated its total lack of enthusiasm to prepare the 477th for duty overseas.

Incredibly, the first complement of officers and men assigned to the 477th were required to make 38 physical moves in the 14 months between the winter of 1944 and the spring of 1945, finally ending up at Freeman Field in Seymour, Indiana. The Army Air Corps appointed as commanding officer Col. Robert Selway, a white man who was an ardent segregationist.

From the outset the black officers and men at Freeman

Field were subject to inhospitable treatment. Most of the restaurants in the town refused to serve any of the black airmen. Some grocery stores refused to sell food to their families. The only laundry in town refused to provide service to the black airmen and their families. Now due to a lack of stockade space in Europe, German and Italian prisoners of war were housed at military bases in the United States, including Freeman Field. These POWs were provided with a fair degree of freedom, and they were served at all the restaurants in town, and the laundry was willing to wash their clothes. They were also often escorted to USO shows, movies and dances at which attendance by the African Americans was prohibited. The irony of this discrimination was a bitter pill for the airmen of the 477th, who nonetheless bore it stoically.

In addition, the African Americans of the 477th were subjected to some incendiary comments from Col. Selway's direct superior, General Frank Hunter, of the First Air Force, who addressed the pilots and crews of the 477th as follows: "This country is not ready or willing to accept a colored officer as the equal of a white one. You are not in the Army to advance your race. Your prime purpose should be in taking your training and fighting for your country and winning the war, and that way you can do a great deal for both your race and your country. As for racial agitators, they shall be weeded out and dealt with."

When the 477th arrived at Freeman Field, there was an officers' club that had been shared by whites and blacks. Once Col. Selway became the commanding officer, he issued an order on April 1, 1945 that restricted the officers' club and certain other buildings to whites only. No black officers were to be admitted.

Two days later, three groups of black officers attempted to enter the whites only officers' club. More than 30 black officers were placed under arrest and were released several days later. On April 9, Col. Selway issued a new regulation stating that strict segregation of base facilities would be enforced. He further ordered all black officers to sign a statement verifying that they read, understood and accepted the conditions of the regulation. Over 100 black officers refused to sign the statement, and they were then advised of the 64th Article of War noting that a court martial would be the consequence of refusing to obey a direct order. On April 12 the 101 African American officers who refused to sign the order were transferred under armed guard to Godman Field, Kentucky and kept under house arrest. News of this quickly became public, and blacks and many whites across America reacted with outrage. A prompt investigation was initiated by the War Department.

Selway was deposed and made the specious argument that his regulation was intended to separate black officers who were in training from the white officers who were their instructors. He stated,

> *The purpose is that during normal training hours the instructor and the student work together . . . and . . . they [should] be given an opportunity during their social hours to forget the instructor-student official attitude, and . . . be permitted to relax from that position of tension from the result of training.*

Selway gratuitously added that he was familiar with "the general discontent always among Negro personnel to be commanded by whites or to be supervised by white personnel." He stated that the Negroes do not desire combat, but they intend to use the military as a vehicle on which to conduct a race crusade.

Fortunately, the charges made their way quickly up to General George Marshall, the Army Chief of Staff, and pursuant to his orders, the 101 mutineers were promptly released. Selway was removed from command and was replaced by none other than Lt. Col. Benjamin Davis. Unfortunately, the continuation of the 477th as a bomber group would be short lived after the surrender of Japan in August.

Of course, the war had not changed the attitudes of most Americans, and the Tuskegee Airmen returned home to the same prejudices and segregationist practices that they had faced before. Nor had the position of the War Department changed regarding strict segregation of blacks and whites. Ironically, the Airmen encountered no such prejudices while deployed in Morocco, Tunisia, Sicily, or Italy, where they were treated as equals along with their white counterparts by the people of those countries.

You would expect that these African American pilots would be looking for jobs in aviation once they returned from the war. Indeed, Eddie Rickenbacker, the President of Eastern Airlines, who had spoken positively during the war to the 99th Fighter Squadron, told them that there will always be places for Americans who work hard. When black pilots applied for jobs at various airlines, including Eastern, following the war, none were hired.

A ray of light appeared in May 1946, when the American Veterans Committee, a largely white group with a great degree of political clout, sent a stinging rebuke to the Army Air Corps and the War Department. It stated,

> *"We feel that the primary lesson learned from World War II is that segregation does not provide maximum efficient utilization of our nation's manpower. [There is strong] evidence of the ability of whites and Negroes to live and work together successfully, as seen in the administration of medical services, Red Cross services, in officers' training schools and on the battlefields when the emergency of the situation necessitated the breaking down of the barriers or racial discrimination . . . . By recommending the continuance of the basic*

> *policy of segregation and only abolishing the most obvious and degrading manifestations of it, the War Department loses a great opportunity and fails to live up to hopes and expectations."*

On July 26, 1948, President Truman signed Executive Order 9981 striking down all segregation in the American military. The order stated in part, "It is hereby declared to be the policy of the President that there should be equality of treatment and opportunity for all persons in the armed services without regard to race, color, religion or national origin."

And on October 27, 1954, Col. Benjamin O. Davis, Jr., commander of the 99th Fighter Squadron and later of the 332nd Fighter Group, and briefly the 477th Bomber Group, was promoted to the rank of Brigadier General. He became the first black American to become a general in the United States Air Force.

So how do we perceive the legacy of the Tuskegee Airmen? They consisted of young black Americans who had a love of flying and who possessed the intelligence and skill to become pilots and ground crews. They were eager to fight for their country, even though their country in many ways had turned its back on them. They were a vanguard of African Americans whose persistence and ultimate success was a large reason for the full integration of the armed services. They proved that a black man could be as much a warrior as his white counterpart. They endured their country's racism and their military's institutionalized segregationist practices stoically. They continued to be patriotic, though they were better treated in Europe, which had no sense of segregation, than they were in their own country. When confronted with gross injustice, as was the case with the Freeman Field Mutiny, they stood their ground on principle. They bore themselves with dignity, and they trained and fought and achieved success as aviators in spite of the institutional barriers of their own military that sought to keep them down. The Tuskegee Airmen must be perceived as an opening wedge in the battle for full desegregation in our country.

A parting comment. In my research for this paper, I was not prepared for the institutionalized racism that existed in the American military. And I was depressingly reminded of the Jim Crowism and segregationist attitudes of northern and southern whites alike. Some of the comments made by the War Department and the Army Air Corps were unbelievable to me in terms of their arrogance, ignorance and sheer cruelty. It's extraordinary what has happened over the past 70 years to wash much of that away. On December 9, 2008, the living Tuskegee Airmen were invited to attend the inauguration of Barak Obama. More than 180 Airmen attended the inauguration. One of them, William Broadwater, 82, made this statement, "The culmination of our efforts and others' was this great prize we were given on November 4th. Now we feel like we've completed our mission."

STATISTICS

- 994 Pilots trained.
- 450 Deployed overseas.
- Over 15,000 combat sorties (including 6,000 plus for the 99th prior to July 1944).
- 111 German airplanes destroyed in the air, another 150 on the ground.
- Approximately 1,000 railcars, trucks, locomotives and other motor vehicles destroyed.
- One destroyer sunk by P-47 machine gun fire.
- 150 pilots killed in action or by accidents.

- 32 pilots downed and captured as POWs.
- 150 Distinguished Flying Crosses earned.
- 744 Air Medals.
- 8 Purple Hearts.
- 14 Bronze Stars.

BIBLIOGRAPHY

Chris Bucholtz, *332nd Fighter Group* - Tuskegee Airmen (2007)

Lyn Homan and Thomas Reilly, *Black Knights: The Story of the Tuskegee Airmen* (2001)

The Round Table Presentation | October 4, 2012

The Mother Of Invention
Bell Laboratories

The title of my paper, The Mother of Invention, may have caused some of you who paid attention to rock and roll in the 60s and 70s, to conclude that this paper was about Frank Zappa's band, the Mothers of Invention. Probably most of you thought of the old adage "necessity is the mother of invention." That would come a bit closer to describing what this paper is about. The mother I am talking about here is Ma Bell, the popular pseudonym for the American Telephone and Telegraph Company, at least before the anti-trust settlement and break up of the Bell system in the 1980's. This paper is about one of the most unique organizations ever established, an organization that, for the greater part of the 20th century, invented the technology that defines much of our contemporary life. This is a paper about Bell Laboratories.

Why is Bell Labs so significant? From its founding in 1925 until its breakup in the 1980s, Bell Labs was hands down the most innovative scientific organization in the world, and was chiefly responsible for making the United States the most technologically advanced nation on the planet.

At its peak in the 1960s, Bell Labs employed more than 15,000 scientists, engineers, mathematicians and technicians. More than 1,200 of its scientists had Ph.Ds. These scientists were responsible for creating thousands of inventions, with the most notable being the transistor, the fundamental building block of every electronic device on the market today. The transistor and the harnessing of atomic energy are generally considered as the two most transformative achievements of the 20th Century.

Seven Nobel Prizes have been awarded to 13 Bell Labs scientists. William Shockley, Walter Brattain and John Bardeen received a Nobel Prize in physics for the invention of the transistor. Three other Bell Labs scientists received the Nobel Prize in physics for the invention in 1969 of the "charge-coupled device," a device that converts light into an electric charge thus enabling digital photography. Steven Chu, the current Secretary of Energy, a scientist at Bell Labs in the 70s and 80s, received the Nobel Prize in physics for developing methods to utilize laser light to cool and slow the motion of atoms. As early as the 1940s, Bell Labs explored the use of microwave technology in telecommunications. In the 1960s, two Bell Labs scientists, Arno Penzias and Robert Wilson, working with microwave technology, discovered cosmic microwave background radiation. This discovery provided persuasive support for, if not confirmation of, the big bang theory, and for that discovery they were awarded the Nobel Prize in physics.

Beyond those discoveries that merited the awarding of Nobel Prizes, scientists at Bell Labs achieved other game-changing innovations. One of those was the laser. The laser was first described, theoretically, in a technical paper submitted by two Bell Labs scientists in 1958. Within two years, scientists and engineers at Bell Labs had created the first gas laser. Of course, the laser has many uses in industry and medicine as well as in electronics. Today if you were to buy a DVD player, the laser that reads the digital information on your disk is smaller than a grain of rice. Lasers used in computers today are many times smaller than a single grain of sand.

In 1954, scientists at Bell Labs created the first working photovoltaic cell, converting the sun's energy into electricity. Within just a few years the improvements made by Bell Labs to the solar cell made it a practical alternative to other forms of electrical power. Bell Labs was also responsible for the first telecommunications satellite known as Telstar. Fittingly, it was powered by solar cells and used transistors to amplify microwave signals it received. Telstar was developed by Bell Labs as an alternative to undersea telephone cables, and was neither sponsored by nor controlled by the U.S. government. Launched from a NASA rocket in July 1962, it transmitted microwave signals across the oceans. Immediately thereafter, concerned about the privatization of satellite telecommunications during the cold war, Congress acted to bring all satellite telecommunications under U.S. government control.

As early as 1947, a Bell Labs paper was the first to propose a network of small interlocking cellular sites using microwave transmission technology that would track telephone users as they move and would pass their calls from one site to another without dropping the connection. Scientists at Bell Labs then created the cell phone technology and in 1978, Bell Labs installed the first commercial cellular network in Chicago.

In 1962, Bell Labs developed the first digital transmission of multiple voice signals. As demand grew to transmit more voice and data communications over the Bell System's network, Bell Labs was the first to explore the use of pulses of light to transmit digital communications. As a result, the Bell Labs scientists pioneered fiber optics technology and actually established the first fiber optic network experimentally in Atlanta in 1976.

The Unix operating system was developed at Bell Labs, and it made large-scale networking of diverse computing systems, and ultimately the Internet, practical. Today Unix is the operating system of most large Internet servers, as well as business and university systems.

The list goes on and on – three other Nobel Prizes in physics for discoveries in the realm of quantum physics and creation of the fax machine which enabled the transmission of not just voice but now data over phone lines. These and literally thousands of other scientific breakthroughs were made possible because Bell Labs had the full financial support of AT&T which, from the early years of the 20th Century, was by far the largest corporation in the world and had the financial wherewithal to engage the nation's finest scientists and engineers and to provide them with the time and tools needed to develop their ideas and inventions. The inventions and innovations at Bell Labs make for a host of interesting stories. I am going to briefly relate two of them, one being the discovery of the transistor, and the other being the creation of information theory which laid the basis for the development of the modern computer. Before I get to those stories, however, let me provide you with some background regarding AT&T and the Bell System.

Alexander Graham Bell invented the telephone in 1874, and earned patents on his invention in 1876 and 1877. Backed by two investors, Bell formed the Bell Telephone Company in 1877, and in 1878 opened the first telephone exchange in New Haven, Connecticut. Within three years, telephone exchanges established by Bell Telephone existed in most major cities in the United States. In 1882, the Bell Telephone Company acquired a controlling interest in the Western Electric Company, which became its manufacturing unit. The entity we know now as AT&T was incorporated in 1885, first as a wholly-owned subsidiary of Bell Telephone. Its original purpose was to build and operate a long distance telephone network. In 1899 a reorganization occurred and AT&T became the parent company of the entire Bell System.

Prior to Bell's second patent expiring in 1894, only Bell

Telephone and its licensees could legally operate telephone exchanges or systems in the United States. But once that second patent expired, within a 10-year period over 6,000 independent telephone companies were in business throughout the United States. Many cities had a half dozen or more separate telephone companies, and this presented a new set of problems. There was no interconnection. Subscribers to different telephone companies could not call each other. In 1907, AT&T's president, Theodore Vail, proposed to Congress that the Bell System function as a legally sanctioned, regulated monopoly. His point was that the telephone, by the nature of its technology, would operate most efficiently if it were a monopoly that could provide universal service. In return, he was willing to accept government regulation as a substitute for the competitive marketplace. AT&T had been buying up many of these independent telephone companies over the prior decade, and it agreed to stop further acquisitions and to connect the remaining independent telephone companies to its network. In 1913, the United States government effectively adopted AT&T's proposal that it become a monopoly regulated by the federal government. As a result, AT&T then owned the principal manufacturer of all telephone equipment in the form of Western Electric, it owned 90% of all operating telephone companies throughout the United States (which became the baby Bells), and it had complete control of all long distance service through its Long Lines division.

The notion of necessity played a central role in the efforts of the scientists at Bell Labs. First and foremost, the laboratory was expected to address the immediate and long range challenges associated with creating, sustaining and improving a national telecommunications network. AT&T's need for a large group of scientists, engineers and technicians was obvious. The Bell System was perceived as a gigantic machine containing millions of component parts. Literally everything that was part of the system had to be invented, assessed and improved upon. The system had to work so that millions of local and long distance calls could be handled simultaneously with distinctive voice transmission travelling through countless miles of wires and switching stations with the assurance that the call would not be dropped or lost or sent to the wrong recipient. For one person to call another required the interrelated and proper functioning of thousands of mechanical and electronic elements. All the elements of the system had to be tested for durability. Forty years was the labs' standard. Teams of chemists had to deal with the creation of proper insulation for the telephone wires. Electronic engineers had to deal with issues such as distortion, feedback, and signal strength. The system had to keep track of the millions of daily calls for billing purposes. Indeed, the issues were so numerous and consequential that a well-staffed scientific laboratory was an essential part of the Bell System.

On January 1, 1925, AT&T officially created Bell Telephone Laboratories as a standalone company housed initially in a 600,000 square foot building in west Greenwich Village. It replaced what had been part of Western Electric's engineering and science department. Bell Labs would be expected to develop new equipment and new technologies for Western Electric and the Bell System and to investigate anything remotely related to human telecommunications.

Much of the long-term success of Bell Labs can be attributable to one man whose name you have probably never heard – Mervin Kelly. Born in rural Missouri in 1894, he was educated as a physicist at the University of Chicago. He had been at AT&T for several years prior to the establishment of Bell Labs. At Bell Labs he quickly moved into the position of Director of Research and ultimately became Chairman of the Board of Bell Labs. Bright and superenergetic, he guided the research at Bell Labs for over 30 years.

Kelly divided the entire scientific organization into three groups. The first was research, creating new inventions and discoveries necessary to improve the system. The second group was systems engineering, where engineers surveyed the new inventions and discoveries and figured out how to integrate them into the existing phone system. The third group comprised the engineers who actually designed and developed these new devices, switches and systems. The movement was from discovery, to development, to manufacture.

Mervin Kelly has long been credited with developing the extraordinary culture of innovation and collaboration among scientists and engineers at Bell Labs. Although there were always problems that required immediate attention, Kelly urged many of the scientists to take the long view, to look out ten to twenty years and to imagine what needs should be addressed. He wanted researchers who would follow their own instincts and explore matters that specifically interested them, not merely what might improve AT&T's bottom line over the short term. As one of the scientists who worked on the development of the silicon transistor stated, "The only pressure at Bell Labs was to do work that was good enough to be published or patented."

Kelly gave his scientists a lot of running room. One writer stated that "Some of Kelly's scientists had so much autonomy that he was mostly unaware of their progress until years after he authorized their work. When he set up the team of researchers to work on what became the transistor, for instance, more than two years passed before the invention occurred. Afterward, when Kelly set up another team to handle the invention's mass manufacture, he dropped the assignment into the lap of an engineer and instructed him to come up with a plan. He told the engineer he was going to Europe in the meantime."

One of the ways Kelly fostered a culture of collaboration was through his design of a new laboratory for Bell Labs. In the late 1930s, he persuaded the AT&T executives that Bell Labs needed a new location and a new structure to house many of its scientists, engineers and technicians. In 1942, Bell Labs moved from its Greenwich Village location to Murray Hill, New Jersey, where Kelly had acquired a 225-acre campus on which he built a huge laboratory that housed a great many of the employees of Bell Labs. In particular, Kelly designed the building so that its scientists and engineers were always in one another's way. He created two four-story wings that had central halls more than 700 feet in length. Standing at one end of a hallway you could literally see the vanishing point at the other end. Kelly insisted that the doors to each of the laboratories along the halls be kept open, and each of the scientists was encouraged to stop and talk to others along the hall. It was generally agreed that chance encounters fostered creative ideas.

One of the most significant challenges to the entire telecommunications system of AT&T was that of amplification. Because of the resistance that occurs in copper wires, an electronic signal containing voice communication that must travel over a long distance needs to be amplified at periodic intervals to allow the signal to reach its destination with sufficient signal strength to be fully heard and understood at that point. In the early 1900s, an inventor named Lee deForest invented a vacuum tube that allowed amplification of electronic signals. AT&T bought the patent and made these vacuum tubes a key part of AT&T's telephone service. However, they were extremely difficult to make, required a great deal of power to operate, gave off enormous amounts of heat, and were often unreliable. A second challenging problem for the Bell System related to the mechanical switches that were employed throughout the system to channel individual telephone calls. These mechanical switches were cumbersome and often unreliable.

In 1945, as scientists came back from the war effort to rejoin Bell Labs, Kelly assembled a team to explore an alternative to vacuum tubes. Leading the team was a brilliant young physicist named William Shockley, a graduate of Cal Tech and MIT. Shockley then drafted Walter Brattain, an experimental physicist at Bell Labs, and John Bardeen, a theoretical physicist from Princeton University, as key members of his team. The focus of the team was upon a particular branch of science known as "solid-state physics." This field addressed the properties of solids, their magnetism and conductivity, in terms of what happens not only on their surfaces, but deep in their atomic structure. In particular, the team concentrated on a certain class of materials known as semiconductors - so named because they are neither good conductors of electricity (like copper), nor good insulators of electricity (like glass). The team was interested in semiconductors such as silicon and germanium because they have a lattice type molecular structure, and exhibit some unique behaviors in terms of atomic structure and the movement of electrons at different levels within a particular slice of semiconductor material. In addition, they are considered good "rectifiers." A rectifier is a substance that will allow electric current passing through it to move in only one direction, thus effectively allowing alternating current to be transformed into direct current. Shockley brought an in-depth knowledge of quantum physics to the team's exploration of the qualities of semiconductors.

Shortly after the team was formed, Shockley designed what he hoped would be the first semiconductor amplifier. Theoretically it should have worked, but it did not, and Shockley assigned Bardeen and Brattain to find out why. For the next two years, Bardeen and Brattain, with occasional input from Shockley, experimented on various types of semiconductors. Then in December 1947, a break-through occurred. Bardeen had an insight as to the way electrons moved below the surface in semiconductors, and as a result of this insight, on December 23, 1947, Bardeen and Brattain produced a device consisting of a combination of silicon and germanium, about the size of a kernel of corn, that produced a signal 18 times more powerful than the electronic signal it received. The first solid-state amplifier had just been invented. The device had no name, but a vote of the team favored the term "transistor," which combined the ideas of transformation and resistance.

When Shockley learned that Bardeen and Brattain had created an effective amplifier from semiconductor material, he was deeply disappointed that he had not been more closely involved in their efforts, and became worried that he would not share in the credit for the development of the transistor. In a combination of anger and despair, Shockley shut himself away for a five-day period and furiously developed a theoretical model for a different type of transistor. It turned out that Shockley's theoretical model was a clear improvement on the one developed by Bardeen and Brattain, though it took two more years to construct a working model. Significantly, Shockley refused to share this work with the rest of the team and as a consequence, the close collaboration of the group dissolved. Shortly thereafter, both Bardeen and Brattain left Bell Labs. As a side note, Bardeen went to the University of Illinois where he explored the super conductivity of materials at low temperatures and for which he received a second Nobel Prize in physics in 1990, making Bardeen the only scientist who received two different Nobel Prizes for physics.

AT&T recognized right away that the transistor was too important a device to keep to themselves, and any effort to do so would probably have been challenged by the government and the courts. So Bell Labs licensed the technology to numerous companies and hosted many seminars to assure the diffusion of their semiconductor technology during the 1950s to a broad range of people, including Jack Kilby at Texas Instruments and Robert Noyce at Fairchild Semiconductor who teamed up to develop the integrated circuit for which they won the Nobel Prize. The integrated circuit is now what we refer to as the computer chip.

The transistor was an amazing new device. It utilized approximately 1 millionth of the power needed to operate a comparable vacuum tube. With new designs the amplification potential increased, and it was also determined that a transistor could be a very effective switching device, and thus could be used to replace the millions of mechanical switches that were needed throughout the Bell System. And, as certain farsighted scientists perceived at the time, the ability of the transistor to serve as a switching device made it the fundamental building block in all computers. A solid-state switch would have no moving parts, making it smaller, faster, quieter and more reliable than the mechanical switches that were then being used.

Over the years, the number of transistors and the amount of information on a computer chip has increased to an extent that boggles the mind. Gordon Moore, a Bell Labs scientist who went on to co-found Intel Corporation, made an observation in the 1960s about the rapid rate of progress in semiconductor engineering. He suggested that the number of transistors that could fit on a silicon chip tends to double about every two years. This is known as Moore's Law, not a law in any scientific sense but an observable phenomenon. As a consequence, today's computer chip is about one-tenth the size of the fingernail on my baby finger, and today it can hold nearly a billion transistors. I read in a recent Scientific American article that scientists have discovered certain properties in a derivative of graphite called graphene which has a lattice-like structure at the molecular level. A transistor made of graphene would be no more than four to eight atoms wide, ten to twelve atoms in length and one atom thick. This must certainly be the ultimate in nanotechnology and most certainly marks the limit of Moore's Law.

Shockley, Bardeen and Brattain received the Nobel Prize for physics in 1956 for the invention of the transistor, but Brattain and Bardeen no longer had any personal or professional relationship with Shockley. Shockley himself left Bell Labs in 1955 to set up Shockley Semiconductor in Palo Alto, California. He took with him a number of physicists and engineers from Bell Labs and is generally thought responsible for the beginning of Silicon Valley. Shockley was a terrible manager and within two years, eight of the scientists who followed him to Palo Alto left the company. Two of those scientists founded Intel Corporation and went on to become billionaires. Shockley Semiconductor did not survive and folded its tent within about two years. Shockley went on to take a teaching post at Stanford. After that, Shockley's life spun out of control and he descended into paranoia and tragedy. He became an outspoken proponent of eugenics, a pseudo science that had been completely repudiated after the horrors of Nazi Germany.

Promptly following the invention of the transistor, a most unexpected and serendipitous event occurred. A mathematician at Bell Labs named Claude Elwood Shannon was introduced to the transistor. It was said that in 1948 when the transistor was announced to the world, you either understood what it could do or you didn't. Claude Shannon understood what it could do. Shannon was viewed as one of the most exceptional minds ever to have graced the premises of Bell Labs. As a graduate student at MIT, Shannon took up flying. An effort was made by the faculty at MIT to dissuade him from continuing because they felt that his mind was too valuable to lose to a flying accident. At MIT, Shannon had worked with Vannevar Bush on a computer that Bush had created to solve differential equations. This gave Shannon early, in-depth experience with these primitive computers. His work caused him to think about how circuits within computers should be constructed. He found he could make sense of the circuits within a computer if he applied the principles of Boolean algebra, an obscure branch of mathematics based on zeros and ones. Shannon wrote up his insights outlining how logic circuits could be designed in a computer. This slender and highly mathematical paper, no more than 25 pages in length, has become known as the most influential masters thesis in history. It described the structure of today's modern computer.

In 1948, Shannon quickly perceived that the transistor could become central to computer design. The transistor became a central piece of a paper he wrote entitled "The Mathematical Theory of Communication." This paper was heralded by Scientific American as the magna carta of the information age. It wasn't about one particular idea, it was about information theory, that is, it set out general rules and unifying ideas for the creation of, and use of, computers. Shannon's colleagues at Bell Labs described his work as "one of the great intellectual achievements of the 20th century," and many compared his genius to that of Einstein. It was commented that if Shockley, Brattain and Bardeen had not discovered the transistor in 1947, somebody else would have discovered it within a few years. But Shannon's thinking was 50 years ahead of anybody else in the area of information theory.

In particular, Shannon perceived the future of telecommunications was not only to transmit the human voice but that ultimately, the system would be used to transmit data from one point to another. And it was essential that if information were to be transmitted over such a system, information introduced into the system at one point must reach the end point without error, in exactly the form in which it was introduced. Shannon established effective strategies to insure the absolute accuracy of the transmission of information. He also perceived that information should be measured in something which he called binary digits or bits, and because of his understanding and appreciation of Boolean algebra (the ones and zeros of computer language), Shannon declared that information should be expressed in a digital format and not otherwise.

Shannon did not win a Nobel Prize for his work, but that is only because there is no Nobel Prize for mathematics. But there is no doubt that Claude Shannon should be credited with creating the template for the modern computer.

As an aside, Shannon had worked on encryption issues and code breaking during the war. In 1949 he published a paper entitled "Communication Theory of Secrecy Systems," considered the foundational treatment of modern cryptography. The paper was immediately classified by the government. In the paper Shannon proved mathematically how one could create an absolutely unbreakable code.

On a personal side, Shannon was one of the more interesting figures ever to work at Bell Labs. He loved to juggle, and could juggle up to 5 balls, which is apparently the dividing line between good and great jugglers. He often rode a unicycle up and down the hallways at Bell Labs juggling as he went. Early in his career, he encouraged the creation of a computer that could play chess against a human. He thought that a game playing computer would unlock secrets as to how humans reason. He was heralded in Time Magazine for the creation of a robotic mouse named Theseus who had copper wire whiskers and a magnet inside his wooden body, and who could navigate a fairly complex maze that Shannon created. The interesting thing about Theseus was that as he bumped his way back and forth around the maze until he found his way out, he was able to learn from his mistakes. His first time through the maze (and the maze could be reconfigured in a myriad of different ways), it may take Theseus several minutes to negotiate his way out. The second time, however, it may take less than a minute and the third time, just a few seconds. Time Magazine said that "Theseus the mouse is smarter than Theseus the Greek," who had to unravel a ball of string to allow him to find his way out of the minotaur's labyrinth. Always whimsical, Shannon created a computer program called "Throback," which did mathematical calculations in Roman Numerals.

Although Bell Labs continued to flourish, by the 1970s the

antitrust division of the Department of Justice concluded that the AT&T monopoly had survived for long enough. In 1974 it initiated a lawsuit to separate the operating companies from AT&T, and on January 1, 1984, pursuant to a settlement that had been reached in 1982, the Bell System breakup officially went into effect. AT&T and Western Electric continued as one combined company which was severed from all the local phone companies. Most Bell Labs employees stayed with AT&T, but about 10% left to serve the needs of the new baby Bells. AT&T then spun off Western Electric into a new company called Lucent Technologies, and many of the remaining Bell Labs employees went to Lucent. For a few years, Lucent did well and was highly valued, but by 2000 it was realized that Lucent's future, which depended on even greater demand for telecommunications switching and transmission equipment, would not come to pass. The rise of the cell phone took care of that. Its stock price dropped from $84 a share to less than $2, and it slashed tens of thousands of jobs, including thousands within Bell Labs. In 2005, having failed to rebound as hoped, Lucent merged with the French telecommunications company Alcatel. What used to be Bell Labs was now less than a third of its former size. By July 2008, only four scientists remained at Alcatel in basic physics research. And in August of 2008, Alcatel announced that it was pulling out of basic science, material physics and semi-conductor research. Bell Labs had been effectively dismantled.

The legacy of Bell Labs is undeniably a rich one. Its extraordinary achievements are one of the primary reasons that the 20th century has been called the American century. Can there be a successor? Not, I believe, in the telecommunications and computer fields, since that is the legacy of Bell Labs. Perhaps there could be a Bell Labs clone that would address the challenges in healthcare or energy or the environment. But where might we find the private corporations with the financial strength and the commitment to R&D to fund another Bell Labs? Perhaps the funding role has now shifted to our federal government, and it may well be that the new model is that the great universities, like Vanderbilt, funded with government grants, will carry on the type of R&D that AT&T funded at Bell Labs. One thing is certain, the world has never seen the likes of Bell Labs which, for over 60 years, was truly the mother of invention.

Sources:

The Idea Factory – Bell Labs and the Great Age of American Innovation, by Jon Gertner (Penguin Press 2012).

Crystal Fire – the Birth of the Information Age, by Michael Riordan and Lillian Hoddesson (W. W. Norton & Company 1997).

Scientific American, July 2012, p. 73

The Round Table Presentation | March 1, 2018

Adventures of a Curious Genius
Richard Feynman

Recently I have enjoyed biographies of Einstein and Leonardo de Vinci, both acknowledged geniuses. Tonight I want to introduce you to my favorite candidate for that distinction. His name is Richard Feynman.

Feynman's primary biographer declares that Richard Feynman was not only a genius, but he was a full-blown magician – that is someone who does things that nobody else can do and presents them to you in ways that are completely unexpected. In addition to his brilliance as a physicist and mathematician, which resulted in his being the youngest physicist at Los Alamos, and later the winner of the Nobel prize in physics, when you study his personal life, it is difficult to reconcile his extraordinary range of interests, even obsessions, with his life-long commitment to serious science. He travelled extensively, learned to play the drums and bongos (he managed to learn some extremely difficult rhythms and that led to his public performances with the bongos), led a samba group in Rio during Carnival, taught himself how to crack safes, studied hypnotism, explored sensory deprivation, was a gifted artist, and he explained many of the theretofore mysterious hieroglyphics of the Dresden Codex, the most important surviving manuscript of the Mayan civilization.

Until the end of his life he loved solving puzzles and problems. He always was available to assist other physicists in analyzing their ideas. He had the particular habit of telling the inquiring person not to give him all the information that they had, but allow him to figure out what made up the elements of the problem before he even thought about a solution.

For more than 40 years Feynman was a central figure in the development of quantum physics. Shortly before his death, two autobiographical books were published containing numerous stories of Feynman's interests outside of the strictly scientific realm. The first, Surely You're Joking Mr. Feynman: Adventures of a Curious Character was a New York Times best seller. It was followed by What Do You Care What Other People Think? The anecdotes in these books describe many of Feynman's interests outside the realm of physics, and these anecdotes are told in Feynman's own words. I'm going to treat you to just a handful this evening.

Richard Feynman was born in 1918 in Queens, New York. His father was a Jewish émigré from Russia, and his mother a Jewish émigré from Poland. Like Albert Einstein and Edward Teller, father of the hydrogen bomb, Feynman was a late talker, and by this third birthday had yet to utter a single word. But when he did finally speak it was with a thick Brooklyn accent which he retained all of his life. Indeed, it was so thick that it was often perceived to be an affectation, and many of his physicist friends commented that Feynman sometimes spoke like a "bum."

At an early age Feynman's family moved from Queens to Far Rockaway on Long Island. Feynman's natural curiosity, inventive spirit, and extraordinary intellect caused him to be drawn early to the workings of radios, which had just gained popularity in the 1920s. It was at about age 10 or 11 when he fixed the radio that belonged in a small hotel in Far Rockaway that was owned by Feynman's aunt. She was delighted to recommend the young boy as someone who could fix radios.

Feynman tells the story in his own words this way:

> *The main reason people hired me was . . . they didn't have any money to fix their radios, and they'd hear about this kid who could do it for less, so I'd climb on roofs and fix antennas, and all kinds of stuff. I got a series of lessons of ever increasing difficulty.*
>
> *One job was really sensational. I was working at the time for a printer, and a man who knew that printer knew I was trying to get jobs fixing radios. And so he sent a fellow around to the print shop to pick me up. The guy is obviously poor – his car is a complete wreck – and we go to his house which is in a cheap part of town. On the way, I say, "What's the trouble with the radio?"*
>
> *He says, "When I turn it on it makes a noise, and then after a while the noise stops and everything is all right, but I don't like the noise at the beginning."*
>
> *I think to myself, "What the heck! He hasn't got any money you'd think he could stand a little noise for a while."*
>
> *And all the time, on the way to his house, he's saying things like "Do you know anything about radios? How do you know about radios? You're just a little boy!"*
>
> *He's putting me down the whole way, and I'm thinking "So what's the matter with him? So it makes a little noise."*
>
> *But when we got there I went over to his radio and turned it on. Little noise? My God! No wonder the poor guy couldn't stand it. The thing began to roar and wobble and made a tremendous amount of noise. Then it quieted down and played correctly. So I started to*

think: "How can that happen?"

I started walking back and forth, thinking, and realized that one way it can happen - that all the tubes are heating up in the wrong order.

So the guy says, "What are you doing? You come to fix the radio but you're only walking back and forth!"

I say, "I'm thinking!" Then I said to myself, all right, take the tubes out and reverse the order in the set. So I changed the tubes around, turned the thing on, it's as quiet as a lamb: it waits until it heats up and then plays perfectly – no noise.

When a person has been negative to you and then you do something like that he usually is 100% the other way kind of to compensate. He got me other jobs, and kept telling everybody what a tremendous genius I was saying, "He fixes radios by thinking!" "The whole idea of thinking, to fix a radio – a little boy stops and thinks and figures out how to do it. I never thought that was possible," he'd say.

Feynman's extraordinary capacity to understand complicated mathematics expressed itself early in his life, and by the time he was 15 years old, he had taught himself trigonometry, advanced algebra, analytic geometry, and both differential and integral calculus. While in high school he received the highest score in the United States by a large margin on what was known as the notoriously difficult Putnam Mathematics Competition. His Far Rockaway High School math team, of which he was the head, won the mathematics competition in New York City, and in his senior year of high school he won the New York University math championship, competing against men considerably older and more formally educated than he.

Feynman applied to Columbia University, but was not accepted because of the school's quota on the number of Jews to be admitted. This was one of those dirty little secrets among the Ivy League Universities – they all apparently set quotas for the acceptance of Jewish students. So Feynman attended MIT. As an undergraduate he succeeded in publishing two separate papers in the Physical Review, the most distinguished publication among physicists.

Graduating in 1939 from MIT, he earned a perfect score in physics on the graduate school entrance exam at Princeton, a feat that was unprecedented and has never been matched. It may not seem surprising that Feynman did poorly on the history and English portions of the test. Indeed, his cognitive abilities may have been somewhat lopsided since much of his writing contains numerous misspellings and grammatical errors. Feynman did not care very much about the need to get the language right. He just wanted to get the science right.

The title of Feynman's first book, *Surely You're Joking, Mr. Feynman!* is a response he heard on his first day at Princeton. In the afternoon it was expected that all the graduate students would attend a tea that was offered by the Dean of the Graduate School at Princeton. Feynman describes the event as follows:

> *So the very afternoon I arrived in Princeton I'm going to the Dean's tea, and I don't even know what a "tea" was or why! I had no social abilities whatsoever; I had no experience with this sort of thing. So I come up to the door, and there is Dean Eisenhart greeting the new students: "Oh, you are Mr. Feynman" he says. "We are glad to have you." So that helped a little because he recognized me somehow. I got through the door and there are some ladies, and some girls too. It's all very formal and I'm thinking about where to sit down and should I sit next to this girl or not, or how I should*

> *behave. I hear a voice behind me.*
>
> *"Would you like cream or lemon in your tea, Mr. Feynman?" It's Mrs. Eisenhart, pouring tea. "I'll have both, thank you," still looking for where I'm going to sit, when suddenly I hear "Heh – heh – heh - heh. Surely you're joking, Mr. Feynman."*
>
> *Joking? What the heck did I just say? Then I realized what I had done. So that was my first experience with this tea business.*
>
> *Later on, after I had been at Princeton longer, I got to understand this "Heh – heh – heh – heh." In fact it was at that first tea as I was leaving that I remembered it meant "You are making a social error." Because the next time I heard the same cackle from Mrs. Eisenhart, somebody was kissing her hand as he left.*

Now as a graduate student at Princeton, there was an expectation that each student would present a seminar on a particular scientific area of interest. This seminar was open to students and faculty alike, but often was sparsely attended. Feynman's presentation was on quantum electrodynamics, the interplay among the electron, the electron's recently discovered antiparticle, the positron, and the photon. It turned out this seminar was the beginning point of his work that brought him the Nobel Prize in physics. Three days before the seminar is to be given, Eugene Wigner, a decorated Princeton faculty member, approached Feynman and told him that John Von Neumann, acknowledged as the greatest mathematician in the world, wanted to attend, as did Professor Wolfgang Pauli from Switzerland, and finally Wigner said that Professor Einstein only rarely comes to our weekly seminars, but since your work is so interesting, he's coming too. I must have turned green when I heard this news.

When the time came to give the talk, Feynman recalls, "Here are these monster minds in front of me, waiting! My first technical talk, and I have this audience! I mean they would put me through the ringer! I remember very clearly seeing my hands shaking as they were pulling out my notes from a brown envelope."

But then a miracle occurred . . . the moment I started to think about the physics and I have to concentrate on what I'm explaining, nothing else occupies my mind – I'm completely immune to being nervous. So after I started to go, I just didn't know who was in the room. I was only explaining this idea, that's all.

But then the end of the seminar came. It was time for questions. Pauli, who was sitting next to Einstein, gets up and says, "I do not sink dis teory can be right, because of dis and dis and dis" and he turns to Einstein and says, "Don't you agree Professor Einstein?" Einstein shakes his head and says, "Noooooo." I always thought that was the nicest "no" that I ever heard.

In the spring of 1942 Feynman received his post-graduate degree from Princeton. His biographer described him at that time as follows:

> *This was Richard Feynman nearing the crest of his powers. At age 23 there was no physicist on earth who would match his exuberant command over the many materials of theoretical science. It was not just a facility of mathematics (though it had become clear that the mathematical machinery emerging from his works was beyond any colleague's separate ability). Feynman seemed to possess a frightening ease with the substance behind the equations, like Albert Einstein at the same age.*

Indeed, Robert Oppenheimer himself noted that he

considered Feynman the most brilliant physicist at Los Alamos.

Feynman's scholarship to Princeton required that he could not be married during its term, but he continued to see his high school sweetheart, Arlene Greenbaum, to whom he had proposed marriage and she had accepted. They determined to marry once he had been awarded his PhD. This was despite his knowledge that Arlene was seriously, in fact terminally, ill with tuberculosis which at the time was an incurable disease. She was not expected to live for more than two years. At the marriage ceremony, which was before a city official on Staten Island, Feynman could only kiss Arlene on the cheek. After the ceremony he took her to a hospital near Princeton where he could visit her on weekends.

After receiving his PhD, with the war raging in Europe and in the Pacific, Feynman was invited to join a team of Princeton scientists who were beginning the work on the development of the atomic bomb. Some months later, Robert Oppenheimer extended a specific invitation to Feynman to join the Manhattan Project at Las Alamos.

Robert Oppenheimer was so interested in being sure that Feynman would join the team of physicists at Los Alamos that Feynman remembers receiving a long distance telephone call from Oppenheimer to inform him that he had found a sanatorium in Albuquerque for Arlene. In the spring of 1943 Feynman and Arlene boarded a train for Albuquerque and Los Alamos. Arlene was transferred to the sanatorium in Albuquerque and Richard visited her every weekend.

At Los Alamos, Feynman was assigned to Hans Bethe's theoretical division. He so impressed Bethe that he was quickly made a group leader overseeing issues related to diffusion. This included predicting the strength of the bomb they were creating based upon the amount of fissionable material that would be used to make the bomb. It also included calculations during the assembly process as to how close the assembly of fissionable material needed to be located before it became critical and caused an explosion or a meltdown that would result in the spread of deadly radiation. In connection with his work in calculating the measure of criticality, he was assigned the obligation of overseeing the assembly operations at Oak Ridge, Tennessee, to be sure they were safely dealing with uranium 235. When he arrived he found that they were storing boxes of uranium next to each other in such close proximity that if there were enough of the U235 isotope in those boxes, the entire place could go critical. He learned that no one working at Oak Ridge at the time had the least idea as to what was involved in making an atomic bomb and what the dangers were in the process that was occurring at Oak Ridge. He convinced his superiors that they must disclose to the top brass at Oak Ridge what was happening at Los Alamos and what the dangers were in the operations at Oak Ridge with respect to U235. Ultimately they concurred, and Feynman returned immediately to Oak Ridge and described to the bug-eyed officers running the place the basics of the bomb and the risks that existed in creating the U235 at Oak Ridge.

At Los Alamos he also met and spoke with Nils Bohr, the father of quantum theory and practically a deity among the physicists at Los Alamos. Bohr came several times to the laboratory under the name of Nicholas Baker. The first time he comes he is surrounded by the physicists. The day before he was due to come for his second visit, Feynman gets a telephone call from Nils Bohr's son, Aage, stating that "My father would like to speak to you." "To me? I'm Feynman. I'm just a .." "That's right. We know who you are. Is 8:00 OK?" Some time later, Aage finally told Feynman why his father wanted to speak with him. The first time Bohr was at Los Alamos he said to his son, "Remember the name of that little fellow in the back over there? He's the only guy who's not afraid of me, and will say when I've got a crazy idea. So next time when these guys who say everything is yes, yes,

Dr. Bohr, get that guy and we will talk with him first."

Feynman felt that the scientists at Los Alamos were too lackadaisical about keeping their papers protected. Due to the top secret nature of their work, those responsible for security constantly warned all the scientists to keep their papers in locked drawers and cabinets. Because there was very little else to do during the small amount of free time available, Feynman decided he would use that free time to learn how to crack safes. He began the study, first from books that dealt with safecracking that he received by mail order. Then he began to experiment directly with the safes at Los Alamos.

There are a host of stories that deal with Feynman's safecracking skills. Suffice it to say that he became quite successful at cracking a number of different safes at Los Alamos, oftentimes leaving notes that he had been there much like the Pink Panther. Here, in Feynman's own words, is one of the stories demonstrating his safecracking methodology.

When I was visiting Oak Ridge I had written a report that had to be Okayed by a Colonel that was in his safe . . . he had a much fancier, two-door safe with big handles and thick steel bolts. The great brass doors swung open and he took out my report to read.

Not having had an opportunity to see any really good safes, I said to the Colonel, "Would you mind while you are reading my report if I looked at your safe?"

"Go right ahead" he said, convinced that there was nothing I could do. Well, I looked at the back of one of the solid brass doors and discovered that the combination wheel was connected to a little lock that looked exactly the same as the little lock unit that was in my filing cabinet at Los Alamos. The same company, same little lock, except that there were these impressive rods that moved sideways and appeared to use a bunch of leverage. The whole leverage system it appeared depended on the same little lock that locks filing cabinets, and I had had a fair amount of success in cracking those locks at Los Alamos.

Meanwhile, he was reading the report. When he finished he said, "All right it's fine." He put the report in his safe, grabbed the big handles and swung the great brass doors together. It sounded so good when they closed, but I know it's all psychological, because it's nothing but the same damned little lock.

I couldn't help but needle him a little bit (I always have a thing about military guys, in such wonderful uniforms) so I said "The way you close that safe I get the idea you think things are safe in there."

"Of course."

"The only reason you think they are safe is because civilians call it a "safe."

He got very angry. "What do you mean – it's not safe?"

"A good safecracker could open it in 30 minutes."

"Can you open it in 30 minutes?"

"I said a good safecracker. It would take me about 45."

"Well!" he said. "My wife is waiting at home for me with supper, but I'm going to stay here and watch you and you're going to sit down there and work on that damn thing for 45 minutes and not open it!" He sat down in his big leather chair, put is feet up on his desk and read.

With complete confidence I picked up a chair, carried over

to the safe and sat down in front of it. I began to turn the wheel at random, just to make such action. After about five minutes, which is quite a long time when you are just sitting there and waiting, he lost some patience. "Well, are you making any progress?"

"With a thing like this, you either open it or you don't."

I figured one or two more minutes would be about time, so I began to work in earnest and two minutes later, clink it opened.

The Colonel's jaw dropped and his eyes bugged out.

"Colonel, I said in a serious tone. Let me tell you something about these locks: when the door to the safe is open and the top drawer of the filing cabinet is left open it is very easy for someone to get the combination That's what I did while you were reading my report. Just to demonstrate the danger you should insist that everybody keep their filing cabinet drawers locked while they are working because when they are opened they are very, very vulnerable."

"Yeah, I see what you mean. It's very interesting."

We were on the same side after that.

The Colonel sent a note around to everybody in the plant which said during his last visit was Mr. Feynman at any time in your office, near your office or walking through your office? Some people answered yes. Those who said yes got another note, "Please change the combination of your safe."

One of Feynman's friends at Los Alamos was Claus Fuchs who confessed to being a spy for the Soviet Union in 1950 and was imprisoned. Ironically, when asked who at Los Alamos was the most likely to be a spy, Fuchs speculated that Feynman with his safe-cracking skills and his frequent trips to Albuquerque was a likely candidate.

Feynman continued to visit Arlene on weekends, borrowing Claus Fuchs' automobile to make the drive from Los Alamos to Albuquerque. Despite all efforts, her condition worsened and she died on June 16, 1945. As she lay dying, the Mayo Clinic was running trials of streptomycin, which it turned out had amazing healing powers with respect to tuberculosis. The release of streptomycin for general use occurred in 1947, two years after Arlene died. Feynman did remarry twice, his third marriage to a woman from Yorkshire, England, Gweneth Howarth, was successful and she bore him a son and a daughter.

Following Arlene's death, Feynman returned to work at Los Alamos and buried himself in various projects. He witnessed the trinity test in July 1945 and celebrated the success of the Manhattan Project with most of his colleagues. It was later that he finally became more reflective and anxious about what a destructive device he had helped create.

After Los Alamos, Feynman first took a teaching position at Cornell, but within several years he moved to Cal Tech in Pasadena. Thus began his most productive years. He invented what became known as the Feynman diagrams which allowed a visualization of the complicated mathematical expressions needed to describe the various interaction of particles at the subatomic level. He also developed a theory of "partons" which led to the discovery and understanding of quarks. He published several books, and created the Feynman Lectures on Physics, a three-volume set that became the classic textbook for advanced physics students for more than a generation. And, of course, he formulated what came to be accepted as the correct theory explaining the interaction among photons, electrons and positrons, for which he won the Nobel Prize in physics in 1965.

Feynman tells the story of how he responded to being

notified of his winning the Nobel Prize:

For many years I would look, when the time was coming around to give out the Prize, who might get it. After a while I wasn't even aware of when it was the right "season." I therefore had no idea why someone would be calling me at 3:30 or 4:00 in the morning.

"Professor Feynman?"

"Hey! Why are you bothering me at this time of the morning?"

"I thought you would like to know that you have won the Nobel Prize."

"Yeah, but I'm sleeping! It would have been better if you had called me in the morning," and I hung up.

My wife said, "Who was that?"

"They told me I won the Nobel Prize."

"Oh, Richard, who was it really?" I often kid around and she is so smart that she never gets fooled, but this time I got her."

The phone rings again: "Professor Feynman, have you heard?"

In a disappointed voice I say "Yeah, yeah, yeah."

Now I began to think how can I turn this all off? I don't want any of this. The first thing is to take the telephone off the hook because calls were coming in one right after the other. I tried to go back to sleep but found it was impossible. I went down to the study to think. What am I going to do? I won't accept the Prize. What would happen then? Maybe that's impossible.

And later that morning I was facing a group of journalists, and one of them asks if I could explain what I did to win the Nobel Prize in a minute or so. I said, "Listen, buddy, if I could tell you in a minute what I did it wouldn't be worth the Nobel Prize!"

In the mid-1980s Feynman contracted a difficult form of cancer which he battled for several years before his death. Nonetheless, he had the energy to make a significant contribution to the space program when he was appointed as one of the members of the Rogers Commission which was authorized to investigate the reasons why the Challenger Space Shuttle blew up upon launch in 1986. Ill though he was, he threw himself into the investigation with all the energy he could muster, and that was considerable. In his book, What Do You Care What Other People Think?, Feynman spends a lengthy chapter telling the story of his involvement with the Rogers Commission, which in the end he found very disappointing because he felt the whole process was so political. Feynman suspected and ultimately proved it was the O rings that circled a portion of the booster rockets that had failed. The morning of the launch of the Challenger the temperature in Cape Canaveral was as low as 29 degrees Fahrenheit. In previous launches the temperature on the day of the launch had not gone below 53 degrees Fahrenheit. And the most dramatic moment in the televised hearings of the meeting of the Commission, was when Feynman stole the show by taking a glass of ice water, immersing a portion of an O ring in the ice water and showing the world how the O ring was unable to function properly when it was chilled down to at or near to 32 degrees. It was Feynman's efforts that established the O rings as the correct cause of the disaster.

What disturbed Feynman most about the results of his investigation was his finding that there were wildly different opinions as to the probability of a space shuttle disaster. The engineers who were responsible for the successful operation of the Shuttle concluded that there was a one in one hundred chance that there would be a disaster. Feynman found that the so-called higher ups, or the managers as he referred to them, gave a prediction that there was only a one in one hundred thousand chance of a disaster. Feynman noted that such a risk assessment implied that NASA could launch a space shuttle each day for 300 years and expect to lose only one to a disaster during that entire time. Indeed, Feynman was so disappointed in the formal outcome of the investigation that he insisted that a separate appendix, containing his personal observations on the reliability of the Shuttle, be attached to the report of the Commission. After an intensive effort to remove the appendix, Feynman won that battle, and it contains a thoughtful consideration about the development of the Shuttle and its future. Feynman was appalled at such a staggering difference between what the engineers and management thought the probabilities of a disaster might be.

Richard Feynman died on February 15, 1988 of complications from cancer. He was 69 years old.

Feynman achieved more of a popular fame after his death in 1988, in part due to his two autobiographical books. He was portrayed by Matthew Broderick in the 1996 movie Infinity, and Alan Alda starred as Feynman in a Broadway play, entitled QED. The BBC also created a drama entitled The Challenger with William Hurt playing Feynman.

Feynman's stature among other physicists transcended the contributions he made to the field of physics. He had a colorful, always curious, personality. He was unencumbered by any false dignity, and he despised excessive self-importance. It was commonplace for a physicist to say about his or her young colleagues, he or she is no Feynman but . . . . Feynman dazzled his colleagues by the flashes of inspiration he had, and he had a gift for elucidating those inspired thoughts and theories. He also left us with a rare bit of wisdom about the human condition. Feynman observed that, "The first principle is that you must not fool yourself, and you are the easiest person to fool."

BIBLIOGRAPHY

Genius: The Life and Science of Richard Feynman, by James Gleick (Pantheon Books, 1992)

Surely You're Joking Mr. Feynman: Adventures of a Curious Character (1985)

What Do You Care What Other People Think? Further Adventures of a Curious Character (1988)

The Pleasure of Finding Things Out – the Best Short Works of Richard P. Feynman (1999)

The Round Table Presentation | February 6, 2020

Never Call Retreat
The Marines At The Chosin Reservoir

In 1950, on the Korean Peninsula, an extraordinary battle occurred at the Chosin Reservoir in North Korea. It pitted a single division of Marines against as many as ten divisions of Communist Chinese soldiers, and it was fought in snow covered mountainous terrain under weather conditions that were unimaginably severe. The battle is referred to as the Frozen Chosin, and those Marines who fought there and survived are known as the Chosin Few.

Japan had long exercised a brutal political and military control over the inhabitants of the Korean Peninsula. In 1945, following the surrender of Japan, the Soviet Union and the United States as occupying powers agreed upon a division of Korea at the 38th Parallel. The Soviets installed Kim Il Sung (grandfather to Kim Jong Un) as the leader of North Korea. The United States supported Syngman Rhee as the South Korean head of state.

On June 25, 1950, North Korean forces, equipped and trained by the Soviets, commenced an unprovoked invasion of the South. South Korea was totally unprepared for the invasion, and within a matter of weeks, the North Korean Army had driven the South Korean forces into a small area around the port of Pusan, dubbed the Pusan Perimeter.

Within two days after the invasion commenced, the Security Council of the United Nations met and agreed to support South Korea by a nine to zero vote in the Security Council. Ironically, the Soviet Union boycotted that meeting, angry that the US had refused to recognize Communist China as the rightful member of the Security Council, and instead supported Nationalist China now on the island of Formosa (Taiwan). Had the Soviet Union been present, its veto would have blocked the approval of the United Nations to support the South Korean armed forces.

President Truman and the majority of Americans enthusiastically approved the leadership of the United States together with the United Nations forces, in an effort to turn back the North Koreans. On June 30, the New York Times headline stated: "Democracy Takes Its Stand," praising President Truman's courageous actions and welcoming the execution of the Communist containment policy that Truman had advocated.

When Truman was confronted with the question as to whether this involvement in the Korean conflict required the approval of Congress, he refused to consider it a war, but rather characterized it as a "police action" by the United Nations in which the United States was participating. When the armistice was concluded in 1953, more than 36,000 Americans had given their lives in this police action.

At the time of the invasion of South Korea, General Douglas MacArthur was in this fifth year as the Military Governor of Japan. A five star General, he was Commander in Chief of US Army forces in the Far East, and head of all United Nations troops in Korea. His critics, and there were many, agreed that MacArthur's primary character flaw was that he had a too solemn regard for his own divinity.

In the summer of 1950, MacArthur was confronted with the need to create a strategy to stop the Korean offensive and expel the North Korean Army from the South. MacArthur devised a plan that proposed a landing by the 1st Division of Marines at Inchon, a port on the western coast of Korea located only 30 miles from Seoul. The Joint Chiefs in Washington initially rejected MacArthur's proposal primarily because Inchon experienced one of the most significant tidal differentials in the world. On average there was a 32 foot difference between high and low tide, meaning that any landing by the Marines would have to be perfectly timed and executed. MacArthur argued that the landing at Inchon would achieve complete surprise and allow the UN forces to cut the supply lines of the North Korean Army and roll it up from the rear. His position won the day and what he predicted is precisely what happened. On September 14 the Inchon landings commenced under the watchful eye of General MacArthur. The Marines were almost unopposed and within three weeks the United Nations forces had recaptured Seoul and destroyed the supply lines to the North Korean soldiers who were on the Pusan Perimeter. The North Korean offensive had been crushed. Remnants of the invaders hurried northward to avoid annihilation as a military force.

MacArthur's strategy was so successful that it ignited a worshipful celebration of his genius by the American press. With MacArthur's star shining so brightly, President Truman sought to buttress his own political standing by requesting a meeting with MacArthur, and basking in the glow of the General's celebrity. The two had never met and it was agreed over MacArthur's grumbling that they would meet on October 15 at Wake Island in the Pacific. MacArthur had to travel 1900 miles from Tokyo and Truman four times that distance. The meeting lasted for two hours, and MacArthur excused himself from having lunch with the President saying that he had serious issues to address in Japan. MacArthur came away enraged at the political hacks that he thought Truman and his cabinet represented.

Within six weeks after the Inchon landing, the United Nations and South Korean forces had expelled the North Koreans from the South and now were prepared to advance north of the 38th Parallel. One of the concerns expressed by the Joint Chiefs of Staff was the possibility that the Chinese Communists would take the move into North Korea as a provocation for them to come into the fight. MacArthur was convinced they would not. He argued that the Chinese Communists had only defeated the Nationalists under Chiang Kai-chek within the past year. He did not think they had the ability to so quickly amass and support an army to invade Korea.

MacArthur's plan to subdue and pacify North Korea involved two parallel military maneuvers. The US 8th Army stationed in Seoul would move north on the western side of the mountain range that ran south to north up the center of North Korea. The 1st Marine Division of some 20,000 soldiers would proceed up the eastern side of the Korean Peninsula to the Chosin Reservoir. There they would join up with the 8th Army and proceed to the Yalu River on the border of China and North Korea.

The Chiefs of Staff in Washington approved MacArthur's plan with one important caveat. If there were any evidence that the Communist Chinese had entered the fray, the decision to invade the North would need to be reexamined.

MacArthur appointed General Edmond Almond, his Chief of Staff, to oversee the parallel marches of the 8th Army and the 1st Marine division into North Korea. General Almond was a disciple of General MacArthur and shared the General's opinion that the war was almost over. It was early November and they expected the boys would be home by Christmas. Almond's subordinate, General Oliver P. Smith who was General of the 1st Marine Division, and who was a cautious, by the book, officer who was not a disciple of MacArthur, was more skeptical of the assurances that the

Chinese Communists would not come in to support the North Korean army.

The 1st Marine Division, 20,000 strong, and led by General Smith, made its way up the only north south road on the eastern side of the mountain chain that led to the Chosin Reservoir. The soldiers in the Division, along with more than 1,000 vehicles, including tanks, armored personnel carriers, howitzers, earth moving equipment and the like, stretched out for miles along the single-lane road.

In early November, as his division was in route to the Chosin Reservoir, General Smith received a report from a unit of South Korean troops stating that they had engaged in a fire fight with soldiers they believed to be Chinese Communists, and that there were very many of them. When General Smith relayed this to MacArthur's headquarters, MacArthur's response was, "There is no positive evidence the Chinese Communists as such have entered Korea." But anecdotal stories continued to surface referring to confrontations with Chinese Communist soldiers. MacArthur's intelligence chief stated authoritatively that these Chinese were merely volunteers, acting on their own and probably part of a token force of zealous Communists who came down to help the North Koreans. This was precisely the message that Chairman Mao wanted to convey.

On November 24, the day after Thanksgiving, General MacArthur and his aides flew from Tokyo to the 8th Army headquarters in North Korea on the western side of the peninsula. MacArthur declared again, "The boys will be home by Christmas." Papers in the United States carried headlines that the Home for Christmas campaign would be the end of the war. Following MacArthur's visit on the 24th, rather than flying directly back to Tokyo, MacArthur ordered his pilot, despite his staff's opposition, to fly up to the border of China and North Korea and fly along the Yalu River so MacArthur could see for himself whether there were any Chinese forces amassed at the border. It was a dangerous decision because MacArthur would be vulnerable to anti-aircraft fire as he flew along the Chinese border. MacArthur looked down from 10,000 feet and saw no evidence of any involvement by the Chinese Communist army. So he flew back to Japan convinced that the Chinese Communists posed no threat.

It turned out that MacArthur was no expert in aerial reconnaissance, because by November 24 as many as a quarter million Chinese soldiers had already crossed into North Korea and another half million were amassed along the Manchurian border. Mao had set his trap, and as many as 15 Chinese Divisions would face the 8th Army and the 1st Marines. Mao ordered 10 of those Divisions to focus on the 1st Marines at the Chosin Reservoir. Mao's eldest son had been killed in an air bombardment he attributed to the Marines, and he gave instructions to those Divisions to annihilate the 1st Marine Division.

How did the Chinese Communist Army avoid being spotted by reconnaissance aircraft? For one thing, they were all dressed in white, and they blended in with the snow-covered landscape. Second, they did not move at all by day, finding places to hunker down until nightfall when they would be up and moving. The Chinese did not have the kind of equipment available to the UN forces. They travelled on foot and were expected to live off the land as much as possible, and they did so.

As the Marines approached the Chosin Reservoir, they were confronted with an extreme cold front that swept down from Siberia. Temperatures dropped during the day to below zero degrees Fahrenheit. For the next two weeks the frigid conditions continued, dropping at night to anywhere from 20 to 40 below. And with the wind chill it felt more like 70 to 80 degrees below zero.

On November 27, General Smith completed the deployment

of the 1st Division of Marines. They were in four locations along the 25 mile single lane road from Koto-ri at the southernmost point to Hagaru, the Toktong Pass, and the village of Yudam-ni on the western side of the Chosin Reservoir. General Smith had allocated 8,000 Marines to the Yudam-ni position since that group was intended to be the Marine component of the force that would move up to the Yalu River and end the war. The central village of Hagaru was where General Smith set up his headquarters. Not only was it central to the disposition of the Marines, but it also had enough flat land to allow an airstrip to be constructed to bring in supplies and take out the dead and wounded.

At midnight on November 27/28, the Marines at Yudam-ni and Toktong Pass received the first attacks by soldiers from Communist China. The surviving Marines recalled the oddness of the attack. First they heard a loud speaker reciting in heavily accented English, "Son of a bitch, Marines we kill! Marines you die." This was accompanied by the sound of bugles, whistles and horns. It turned out that the Chinese soldiers, lacking radios, signaled one another through these more primitive means. The single benefit to the Marines was that it gave them a sense of where these Chinese were attacking from. The Chinese came out of the darkness clothed in white and with bayonets fixed. That first night the Marines endured several waves of attacks from midnight until dawn. The sheer number of Chinese soldiers attacking often overwhelmed the Marines and enabled the Chinese to reach the foxholes on the perimeter, necessitating desperate hand to hand fighting during many of the attacks. At the end of the first night at Yudam-ni and Toktong Pass it was determined that more than 1,000 Chinese had died. No one knew for sure, but they did know the Marines had held, although they had lost 50 men killed in the fight and with many wounded. The Marines superior fire power and unwillingness to give ground frustrated all the attacks that first night. The attacks would continue each night for the next 13 days. From darkness to dawn each night the Marines had to be awake and alert for the next wave of Chinese. So the Marines crawled into their sleeping bags to grab some hours of sleep during the daylight.

On November 28, General MacArthur was stunned to learn of the attacks by the Chinese Army against both the 8th Army and the 1st Marine Division. In a matter of four days the circumstances that General MacArthur had guaranteed to be the case, that is that the war would be over by Christmas, were utterly destroyed. The certainty of success just days before had now vanished. MacArthur now was convinced that the war was lost, and he hurried to put the best spin on the situation as he possibly could. He now claimed that he had sent his armies north expressly to test the size and disposition of the Chinese troops. His advances had accomplished everything that he had planned. He claimed to have prodded the Chinese to assault so that he could have them show their hand. MacArthur had not been taken by surprise. His troops did not go blindly into a massive ambush. The nature of the action MacArthur had taken he called "A reconnaissance in force." As far as MacArthur was concerned it had worked perfectly.

The Joint Chiefs of Staff and the President saw through this sophistry immediately. MacArthur had been outwitted and outflanked by an army that had no significant equipment, that travelled on foot, had no tanks and very little artillery, no air force, and primitive communications. In addition to the encirclement of Marines on the eastern side of the mountain range, that same night as many as five Divisions of Communist Chinese soldiers attacked the US 8th Army as it was moving up the western side of North Korea. The 8th Army was totally surprised and after two days of fighting was in full retreat southward. Within a period of four days the situation in Korea had been transformed from a victory for the United States and the United Nations to, in MacArthur's view, a defeat at the hands of the Chinese.

In addition to the relentless nighttime attacks by the Chinese, during daylight hours, though there were no such attacks, there were Chinese snipers all along the perimeter of the defenses. The snipers would fire into the tents, including the hospital tents. They aimed for the truck drivers, the operators of the earth-moving equipment who were preparing the airstrip to receive cargo planes. And the Marines walking around during daylight hours could be easy targets for the snipers as well.

But despite the sniper fire and the countless attacks in the darkness, the greatest combatant that the Marines faced was the cold. Canteens and C-rations froze solid. Flesh stuck to metal. Helicopters refused to rise and truck engines balked. Batteries fizzled. The oil in the machine guns solidified. The Marines could not dig foxholes with their spades because the ground was frozen so solidly. They improvised by taking small pieces of C-4 explosive and literally blowing holes in the ground for their foxholes. The quartermasters had issued the men warm winter clothing – wind-proof trousers, alpaca-lined parkas, heavy woolens, and mountain sleeping bags. But that gear was totally inadequate to combat this kind of deep cold. The cold dulled their senses. Many were dazed and semi-conscious. Their vital signs became erratic and respiratory rates dropped to dangerously low levels. Some were catatonic and overcome by sadness, sobbing uncontrollably. Many said that it was not only their capacity for physical activity that diminished, but even their speed of thought. General Smith himself found it increasingly difficult even to move his jaw to speak. To start an engine required hours of work thawing its moving parts. Blood plasma froze and orderlies were obliged to carry morphine syrettes in their mouths to maintain their fluidity for the wounded men.

The space heaters in warming tents became the very focus of life. During daylight hours, Marines at the perimeter could be spelled and given time to warm up. How the Chinese were able to survive in this weather was a mystery.

One unexpected benefit of the severe cold was that it had a cauterizing effect on wounds. Blood from bullet holes or shrapnel tears simply froze to the skin and stopped flowing.

In the end most casualties suffered by the men were from the extreme cold. If you stopped moving, you froze.

The creation of a workable air strip not only permitted the receipt of needed supplies, but it also allowed several journalists to come to the Chosin Reservoir. The stories they wrote about the nightmarish situation facing the Marines galvanized the American public who collectively waited breathlessly for a successful break-out from the encirclement.

After the first night of attacks from the Communist Chinese, and the recognition that they were encircled by a larger force, there were no longer any thoughts of pushing forward to the Yalu River. Instead, General Smith formulated a plan for a break-out from the encirclement. It involved gathering the troops from Yudam-ni back to Hagaru, bringing in the Toktong Pass soldiers along as well, and then making a march subject to constant attack from Hagaru to Koto-ri. Once at Koto-ri the passage down to the sea involved some engineering feats allowing the Marines to cross a river the bridges over which had been destroyed.

A march from Yudam-ni and Hagaru down to the coast was going to be a challenge. In military parlance is sometimes referred to as a "retrograde maneuver," also known as a "advance to the rear." Therefore it might be viewed as a retreat, although Smith did not allow that term to be used in his presence. When a journalist challenged Smith that this was a retreat, Smith responded, "Retreat hell, we are just attacking in a different direction." Whatever euphemism one wanted to use, all the martial textbooks agreed on this point, even under favorable circumstances, a disciplined, well-executed fighting withdrawal was one of the most difficult maneuvers in military science. It's hard enough for an army to defend itself when dug in, but to do so while it's on the move being attacked relentlessly from all sides for every step of extraction is next to impossible.

The 6th Century BC author of The Art of War, Sun Tzu, describes nine kinds of situations in which an army can find itself. He uses the term ground as a synonym for situation. The ninth and most distressing type of "ground" is one where an army is about to be annihilated and can be saved only by fighting without delay. It is a place that has no shelter nor is there any possibility of an easy retreat. The army has no alternative but to fight its way out of the encirclement or die. Sun Tzu calls this ninth ground "death ground."

David Halberstam, whose last book before his untimely death in a car accident, entitled The Coldest Winter, covers the first year of the Korean War. Halberstam had the following to say about the breakout of the Marines from the encirclement at the Chosin Reservoir:

> *The break out from the Chosin Reservoir is one of the classic moments in [the Marines'] own exceptional history, a masterpiece of leadership on the part of their officers, and a simple, relentless abiding courage on the part of the ordinary fighting men – fighting a vastly larger force in the worst kind of mountainous terrain in unbearable cold that sometimes reached down to minus 40. Of all the battles in the Korean War, it is probably the most celebrated, deservedly so . . . As the news reached Washington and then the Country about the dilemma of the 1st Marines, seemingly cut off and surrounded by a giant force of Chinese, there was widespread fear that the Division might be lost. Omar Bradley himself was almost certain they were lost. When the 1st Marines started the breakout, there were six Chinese divisions aligned against them, roughly 60,000 soldiers, in the two week battle in which the*

> *Marines fought their way back to the sea, General Smith believed they had fought all out against seven Chinese Divisions and parts of three others.*

It is estimated that between 40,000 to 60,000 Chinese were killed during this encirclement and breakout, and perhaps another 20,000 to 40,000 were wounded. Nobody will ever know for sure. From November 27 to December 11 the Marines lost 561 dead, 182 missing, 2,894 wounded and another 3,600 who suffered primarily from frostbite. The frostbite took a heavy toll on toes, fingers, hands and feet.

On April 11, 1951, General Douglas MacArthur was relieved of command by the President. Over the course of the winter, MacArthur had ignored the instructions of the President and the Joint Chiefs of Staff concerning comments that were contrary to the stated position of the United States government. MacArthur advocated an attack on Communist China, and he urged the bombing of Chinese cities, sought a reinvigoration of Chiang Kai-chek, and recommended joining forces with the Nationalists on Formosa against the Communists. I'm sure you remember his recommendation that we place highly radioactive waste along the border between Manchuria and North Korea, thus making it uninhabitable for thousands of years.

Although technically a retreat, military historians regard the battle at the Chosin Reservoir as the defining battle of the modern Marine Corps. It ranks with the conduct of the Marines at Belleau Wood and Iwo Jima. It is used to stress the never give up attitude of the Marines. Never surrender, even in the face of almost certain death.

BIBLIOGRAPHY

David Halberstam, *The Coldest Winter – America and the Korean War* (2007)

Max Hastings, *The Korean War* (1987)

Hampton Sides, *On Desperate Ground* (2019)

Sun Tzu, *The Art of War* (translated by Samuel B. Griffith – 1963)

Made in the USA
Columbia, SC
12 September 2024

6e4e9b6a-8d05-477d-a5aa-1b0fad28ec57R01